Nanomaterials and Nanotechnology in Food: Applicability, Functionality, and Safety

Nanomaterials and Nanotechnology in Food: Applicability, Functionality, and Safety

Guest Editors

Wenxin Zhu
Xiaoyue Yue

MDPI

Basel • Beijing • Wuhan • Barcelona • Belgrade • Novi Sad • Cluj • Manchester

Guest Editors
Wenxin Zhu
College of Food Science and Engineering
Northwest A&F University
Yangling
China

Xiaoyue Yue
College of Food and Biological Engineering
Zhengzhou University of Light Industry
Zhengzhou
China

Editorial Office
MDPI AG
Grosspeteranlage 5
4052 Basel, Switzerland

This is a reprint of the Special Issue, published open access by the journal *Foods* (ISSN 2304-8158), freely accessible at: https://www.mdpi.com/journal/foods/special_issues/nanomaterials_nanotechnology.

For citation purposes, cite each article independently as indicated on the article page online and as indicated below:

Lastname, A.A.; Lastname, B.B. Article Title. *Journal Name* **Year**, *Volume Number*, Page Range.

ISBN 978-3-7258-8277-9 (Hbk)
ISBN 978-3-7258-8278-6 (PDF)
https://doi.org/10.3390/books978-3-7258-8278-6

Contents

Yongxu Wang, Guolin Chen, Daru Wang, Jing Zhang, Chunxiang You, Xiaofei Wang and Huaifeng Liu
Post-Harvest Application of Nanoparticles of Titanium Dioxide (NPs-TiO_2) and Ethylene to Improve the Coloration of Detached Apple Fruit
Reprinted from: *Foods* **2023**, *12*, 3137, https://doi.org/10.3390/foods12163137 . . . . **1**

Jinqiong Xu, Jianbo Liu, Wuwu Li, Yongsheng Wei, Qinglin Sheng and Yonghui Shang
A Dual-Signaling Electrochemical Aptasensor Based on an In-Plane Gold Nanoparticles–Black Phosphorus Heterostructure for the Sensitive Detection of Patulin
Reprinted from: *Foods* **2023**, *12*, 846, https://doi.org/10.3390/foods12040846 . . . . **20**

Shuying Yang, Ruixi Yang, Jingyi He, Yu Zhang, Yahong Yuan, Tianli Yue and Qinglin Sheng
Au Nanoparticles Functionalized Covalent-Organic-Framework-Based Electrochemical Sensor for Sensitive Detection of Ractopamine
Reprinted from: *Foods* **2023**, *12*, 842, https://doi.org/10.3390/foods12040842 . . . . **34**

Keiya Shimamori, Tomohiko Nambu, Daiki Kawamata, Masahiro Kuragano, Naoki Nishishita, Toshifumi Iimori, et al.
Cultivation Factors That Affect Amyloid-β Aggregation Inhibitory Activity in *Perilla frutescens* var. *crispa*
Reprinted from: *Foods* **2023**, *12*, 486, https://doi.org/10.3390/foods12030486 . . . . **48**

Min Li, Pengxiang Wang, Xu Zhang, Hongyu Wang, Ke Li and Yanhong Bai
Development of a Modified QuEChERS Method Based on Magnetic Multi-Walled Carbon Nanotubes as a Clean-Up Adsorbent for the Analysis of Heterocyclic Aromatic Amines in Braised Sauce Beef
Reprinted from: *Foods* **2023**, *12*, 138, https://doi.org/10.3390/foods12010138 . . . . **65**

Xiaoyue Yue, Chaoyun Wu, Zijun Zhou, Long Fu and Yanhong Bai
Fluorescent Sensing of Ciprofloxacin and Chloramphenicol in Milk Samples via Inner Filter Effect and Photoinduced Electron Transfer Based on Nanosized Rod-Shaped Eu-MOF
Reprinted from: *Foods* **2022**, *11*, 3138, https://doi.org/10.3390/foods11193138 . . . . **79**

Biao Zhang, Ruofan Zhou, Huiqi Zhang, Danfeng Cai, Xiaodong Lin, Yihan Lang, et al.
A Smartphone Colorimetric Sensor Based on Pt@Au Nanozyme for Visual and Quantitative Detection of Omethoate
Reprinted from: *Foods* **2022**, *11*, 2900, https://doi.org/10.3390/foods11182900 . . . . **90**

Shujuan Chen, Wanlin Ouyang, Yiting Zhu, Li He, Likou Zou, Xiaolin Ao, et al.
Facile Synthesis of N, S-Doped Carbon Quantum Dots from Food Waste as Fluorescent Probe for Sensitive Detection of Thiamphenicol and Its Analogues in Real Food Samples along with an Application in Bioimaging
Reprinted from: *Foods* **2022**, *11*, 2414, https://doi.org/10.3390/foods11162414 . . . . **100**

Andrea P. Cuevas-Gómez, Berenice González-Magallanes, Izlia J. Arroyo-Maya, Gustavo F. Gutiérrez-López, Maribel Cornejo-Mazón and Humberto Hernández-Sánchez
Squalene-Rich Amaranth Oil Pickering Emulsions Stabilized by Native α-Lactalbumin Nanoparticles
Reprinted from: *Foods* **2022**, *11*, 1998, https://doi.org/10.3390/foods11141998 . . . . **116**

foods

MDPI

Article

Post-Harvest Application of Nanoparticles of Titanium Dioxide ($NPs-TiO_2$) and Ethylene to Improve the Coloration of Detached Apple Fruit

Yongxu Wang [1,2], Guolin Chen [2], Daru Wang [2], Jing Zhang [2], Chunxiang You [2], Xiaofei Wang [2] and Huaifeng Liu [1,*]

1 Xinjiang Production and Construction Corps Key Laboratory of Special Fruits and Vegetables Cultivation Physiology and Germplasm Resources Utilization, Department of Horticulture, College of Agriculture, Shihezi University, Shihezi 832003, China; 15615541161@163.com

2 Apple Technology Innovation Center of Shandong Province, Shandong Collaborative Innovation Center of Fruit & Vegetable Quality and Efficient Production, National Key Laboratory of Wheat Improvement, College of Horticulture Science and Engineering, Shandong Agricultural University, Tai'an 271018, China; 15953327292@163.com (D.W.); zjlucky815@163.com (J.Z.); youchunxiang@126.com (C.Y.); xfwang2004@163.com (X.W.)

* Correspondence: lhf_agr@shzu.edu.cn

Abstract: In this study, we analyzed the effects of treatments with titanium dioxide nanoparticles ($NPs-TiO_2$) and ethylene on anthocyanin biosynthesis and reactive oxygen species (ROS) metabolism during light exposure in ripe 'red delicious' apples. Both treatments led to improved anthocyanins biosynthesis in detached mature apples, while the $NPs-TiO_2$ had less impact on the fruit firmness, TSS, TA, and TSS/TA ratio. Furthermore, the effects of both treatments on the expression of anthocyanin-related enzymes and transcription factors in the apple peel were evaluated at the gene level. The differentially expressed genes induced by the two treatments were highly enriched in the photosynthesis and flavonoid biosynthesis pathways. The expression of structural genes involved in anthocyanin biosynthesis and ethylene biosynthesis was more significantly upregulated in the ethylene treatment group than in the $NPs-TiO_2$ treatment group, and the opposite pattern was observed for the expression of genes encoding transcription factors involved in plant photomorphogenesis pathways. In addition, the ROS levels and antioxidant capacity were higher and the membrane lipid peroxidation level was lower in fruit in the $NPs-TiO_2$ treatment group than in the ethylene treatment group. The results of this study reveal differences in the coloration mechanisms induced by $NPs-TiO_2$ and ethylene in apples, providing new insights into improving the color and quality of fruits.

Keywords: apple coloration; nanoparticle; $NPs-TiO_2$; reactive oxygen species; ethylene

1. Introduction

The apple is an economically important fruit crop that is widely consumed. According to the FAO, the worldwide apple production quantity is about 82 million tons, of which China produces more than 42 million tons [1,2]. Consumers usually associate the red color of apple fruit with ripeness and a sweet taste, thereby establishing red apple cultivars as particularly favored among consumers [3]. The biosynthesis of anthocyanins in the apple peel is affected by a variety of environmental factors and plant hormones [4], with solar illumination being one of the most important environmental factors. Reflective mulch is usually used to promote apple fruit coloration and improve the utilization effect of light [5,6]. In addition, ultraviolet (UV) radiation and low temperatures are particularly important for regulating the production of anthocyanins in apples [7,8].

The typical approaches employed to artificially improve the deposition of pigments in the apple epidermis involve the use of cultivation techniques that facilitate efficient light utilization to promote anthocyanin biosynthesis, as well as the application of chemically synthesized biostimulants or the like to promote ripening and accelerate anthocyanin accumulation [9]. A variety of cultivation strategies have been used to promote the deposition of pigments in apple fruits, such as approaches that increase the transmissibility of light [5,10], micro-sprinkler irrigation systems [11], fruit bagging [12], delayed harvesting [13], and spray applications of chemicals such as exogenous ethylene [14,15]. However, spray applications of chemically synthesized materials promote defoliation, which improves the penetration of sunlight and accelerates the ripening of fruits, which reduces their storage tolerance [16,17]. In addition, consumers might be discouraged to purchase fruits that have been treated with ethylene-releasing compounds because of their potential health hazards [18].

Nanotechnology is expected to provide important alternatives to conventional cultivation methods [19]. Engineered nanomaterials have been widely used in agriculture for improved crop quality and efficiency and to enhance the ability of crops to deal with environmental stress [20,21]. Titanium dioxide nanoparticles (NPs-TiO_2) are some of the most widely used engineering nanomaterials [22]. NPs-TiO_2 are chemically stable, non-toxic, and highly reactive, and have been shown to effectively absorb UV radiation [23]. When NPs-TiO_2 are exposed to UV light with a wavelength equal to or less than 387.5 nm, reactive oxygen species (ROS) can be generated via the photocatalysis of H_2O [24,25]. These photocatalytically generated ROS can attack and degrade organic material (e.g., plant organs, microorganisms, viruses, and organic compounds) adsorbed on the surfaces of NPs-TiO_2 in both aqueous and gaseous phases [26–29]. A previous study has shown that the attachment of NPs-TiO_2 to the surface of the exocarp can accelerate the coloring of the exocarp by absorbing UV light, and the NPs-TiO_2 particles attached to the fruit's epidermal surface can be washed off using a common tap process [30]. In addition, Surround™ WP (95% kaolin) was applied to reduce sunburn on fruit by reflecting more light [31]. The mechanisms by which NPs-TiO_2 treatment regulate anthocyanin metabolism in apple fruit during light exposure remain unclear. Here, differences in the coloring mechanisms of apples and the regulation of anthocyanin biosynthesis under treatments with NPs-TiO_2 and the conventional plant growth regulator ethylene were investigated. The results subsequently revealed differences in the coloring mechanisms of apples induced by NPs-TiO_2 and ethylene, which have implications for the development of new agricultural applications of photocatalytic nanoparticles.

2. Results

2.1. Changes in Peel Color and Fruit Quality during Light Exposure

Figure 1 shows the changes in apple fruit phenotype and skin color during light exposure. The rind coloring process was significantly accelerated in both treatments compared with the control (Figure 1A). The color parameters (Figure 1B–E) measured included the brightness (L^*), redness (a^*), yellowness (b^*), and hue angle (h^*). During light exposure, the a^* value was highest in the ethylene treatment, followed by the NPs-TiO_2 treatment and the control (Figure 1C). The h^* value of the control decreased more slowly than that of the NPs-TiO_2 and ethylene treatments, and there was a significant difference ($p < 0.05$) on the second day of treatment (Figure 1E). The results indicated that the utilization of NPs-TiO_2 and ethylene treatments led to noticeable changes in the color and quality of the fruit epidermis during light exposure.

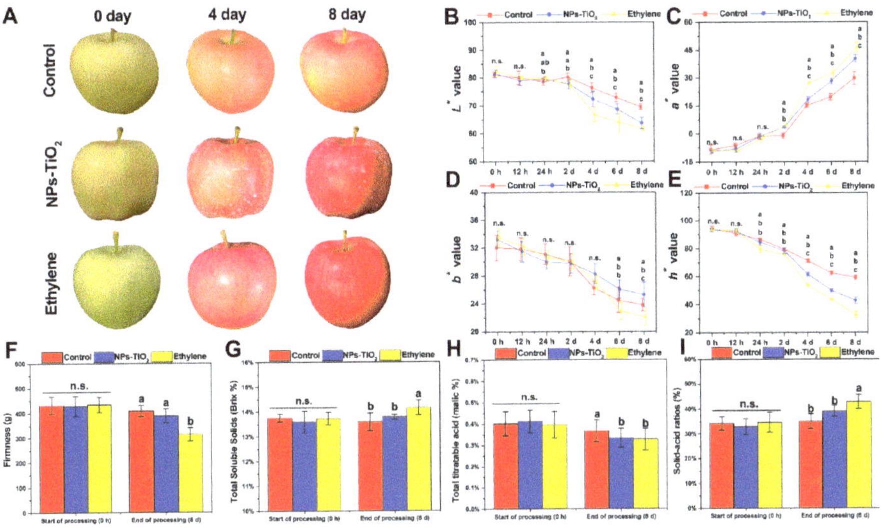

Figure 1. Effects of NPs-TiO_2 and ethylene treatments on color and fruit physiological quality of "Red Delicious" apple during light exposure. (**A**) Fruit coloring phenotype after treatment, (**B**) L^* value (black to white), (**C**) a^* value (green to red), (**D**) b^* value (yellow to blue), (**E**) h^* value (hue angle), (**F**) Firmness, (**G**) Total soluble solids (TSS), (**H**) Total titratable acid (TA), and (**I**) Solid–acid ratios (TSS/TA ratio). Error bars, mean ± standard error ($n = 6$). Different letters indicate significant differences among treatments in the same period using Duncan's test ($p < 0.05$), n.s. represents no significant difference ($p \geq 0.05$).

Fruit quality at the end of the 8-day treatment period was measured. The firmness of fruit was significantly lower in the ethylene treatment ($p < 0.05$) than in the NPs-TiO_2 treatment and control (Figure 1F). The total soluble solids (TSS) (Figure 1G) and solid–acid ratio (Figure 1I) were significantly ($p < 0.05$) higher in the ethylene treatment than in the NPs-TiO_2 treatment and the control; however, there were no significant differences ($p \geq 0.05$) between the NPs-TiO_2 treatment and the control. In addition, the fruit respiration rate (Figure S2A) and ethylene production (Figure S2B) were significantly higher ($p < 0.05$) in the ethylene treatment than in the other two treatments at 8 days. These findings indicate that the fruits in the ethylene treatment were more mature after they were harvested at the end of the in vitro light culture.

2.2. *Effects of NPs-TiO_2 and Ethylene Treatments on the Secondary Metabolites in Exocarp Tissue*

The color of the skin is closely related to pigment deposition, and the most common pigments in the skin of apple fruits are anthocyanins. Flavonoids and phenolic compounds can also affect the coloration of apple fruits [32]. The content of anthocyanins, proanthocyanins, flavonoids, and total phenols was determined during light exposure (Figure 2). The anthocyanin and proanthocyanin content of the exocarp treated with the NPs-TiO_2 and ethylene treatments significantly differed ($p < 0.05$) from that in the control on the second day of treatment (Figure 2A,B). After 8 days of light exposure, the anthocyanin content of fruits sprayed with ethephon reached 293.44 ± 18.01 μg g^{-1}, which was 1.11 times and 1.46 times higher than that in the NPs-TiO_2 treatment and the control, respectively (Figure 2A). Variation in the content of flavonoids and anthocyanins among treatments was similar, and the content of flavonoids and anthocyanins was significantly higher in the ethylene treatment ($p < 0.05$) than in the NPs-TiO_2 treatment and the control on the sixth day (Figure 2C). The content of total phenolics in the fruit epidermis continued to increase over time, and the content measured in the NPs-TiO_2 treatment group was significantly higher ($p < 0.05$) than that of the remaining two treatments on the sixth day. On the eighth day of treatment, the total phenolics content of fruit in NPs-TiO_2 group finally reached 1661.17 ± 77.87 μg g^{-1}, which was 1.10 times and 1.24 times of ethylene and control, respectively (Figure 2D).

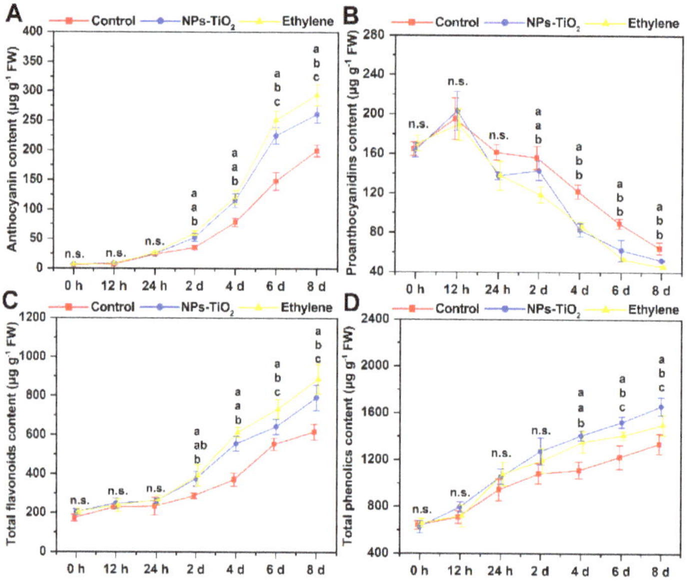

Figure 2. Effects of two treatments on (**A**) anthocyanin content, (**B**) proanthocyanidin content, (**C**) total flavonoids content and (**D**) total phenolics content during light cultivation of 'Red Delicious' apples. Error bars, mean ± standard error ($n = 3$). Different letters indicate significant differences among treatments in the same period using Duncan's test ($p < 0.05$), n.s. represents no significant difference ($p \geq 0.05$).

2.3. *Transcriptome Analysis of the Differently Expressed Genes (DEGs)*

To identify DEGs between treatment groups, transcriptomic analyses were conducted using samples collected during 8 days of treatment. Correlation analysis of the data in each group, principal component analysis scores, and a Venn diagram of DEGs between treatments revealed that the transcriptome data were reliable (Figure 3A,B). Venn diagram showed that 1124 DEGs were overlapped between clusters of control vs. NPs-TiO_2 and control vs. ethylene, which was much higher than the number of DEGs compared between the other two groups (Figure 3C). A volcano plot was constructed to clarify the distribution of DEGs among the control, NPs-TiO_2 treatment, and ethylene treatment ($\log_2$(Fold Change) > 1, q-value < 0.05). The results revealed 3416 DEGs between the control and NPs-TiO_2 treatment, including 2298 up-regulated DEGs and 1118 down-regulated DEGs (Figure 3D). A total of 1719 DEGs were detected between the control and ethylene treatment, including 1264 up-regulated DEGs and 455 down-regulated DEGs (Figure 3E). A total of 934 genes were detected between the NPs-TiO_2 and ethylene treatments, including 379 up-regulated DEGs and 555 down-regulated DEGs (Figure 3F).

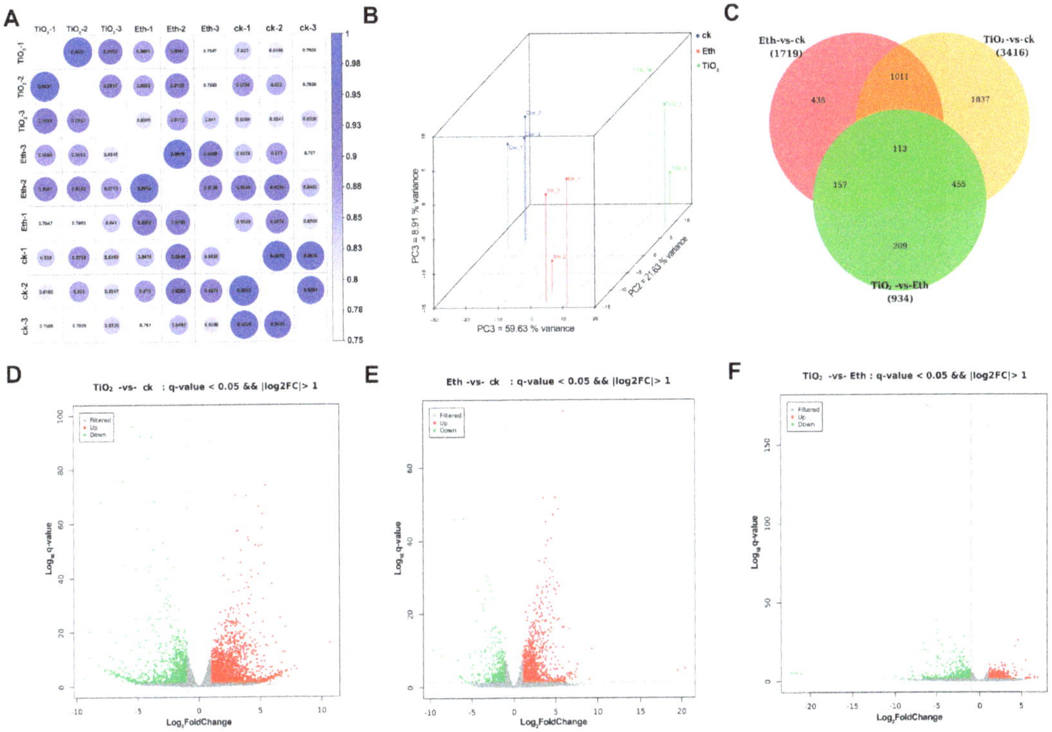

Figure 3. Differentially expressed genes (DEG) in apple after NPs-TiO_2 and ethylene treatment. (**A**) Correlation heatmap of different treatment groups, (**B**) PCA score diagram of DEGs, (**C**) Venn diagram showing the shared and distinct DEGs among the three compared groups, (**D–F**) Volcano plots displaying the up-regulated and down-regulated genes between different treatment groups. ck represents Control, TiO_2 represents NPs-TiO_2, and Eth represents Ethylene.

2.4. Gene Ontology (GO) and Kyoto Encyclopedia of Genes and Genomes (KEGG) Pathway Enrichment Analyses of the DEGs

GO analysis of 2143 (Control vs. NPs-TiO_2), 1493 (Control vs. Ethylene), and 1153 (NPs-TiO_2 vs. Ethylene) DEGs revealed that the 30 most significantly enriched GO terms were in three categories: biological process, cell component, and molecular function. DEGs in the Control and the two treatments encoded enzymes with high catalytic activity, and they were mainly enriched in 'chloroplast thylakoid membrane', 'thylakoid and chloroplast', and 'chloroplast' (Figure 4A,B). DEGs in the NPs-TiO_2 vs. Ethylene comparison group were mainly enriched in GO terms in the biological process and molecular function categories; these DEGs were also enriched in several GO terms in the cellular component category. Within the molecular function category, the most highly enriched terms were 'UDP-glucosyltransferase activity' and 'enoyl-[acyl-carrier-protein] reductase (NADH) activity' (Figure 4C).

Next, KEGG pathway enrichment analysis was performed on selected DEGs. DEGs in the Control vs. NPs-TiO_2 and Control vs. Ethylene comparison groups were significantly enriched in the 'photosynthesis' and 'flavonoid biosynthesis' pathways (Figure 4D–F). We performed real-time quantitative polymerase chain reaction (qRT-PCR) and transcription fragment per kilobase/million fragment profiling to validate the expression patterns of 12 candidate DEGs enriched in the above KEGG signaling pathways (Figure S3). The expression patterns of nine of the 12 DEGs inferred by the qRT-PCR analyses were consistent with those inferred by the transcriptome analysis.

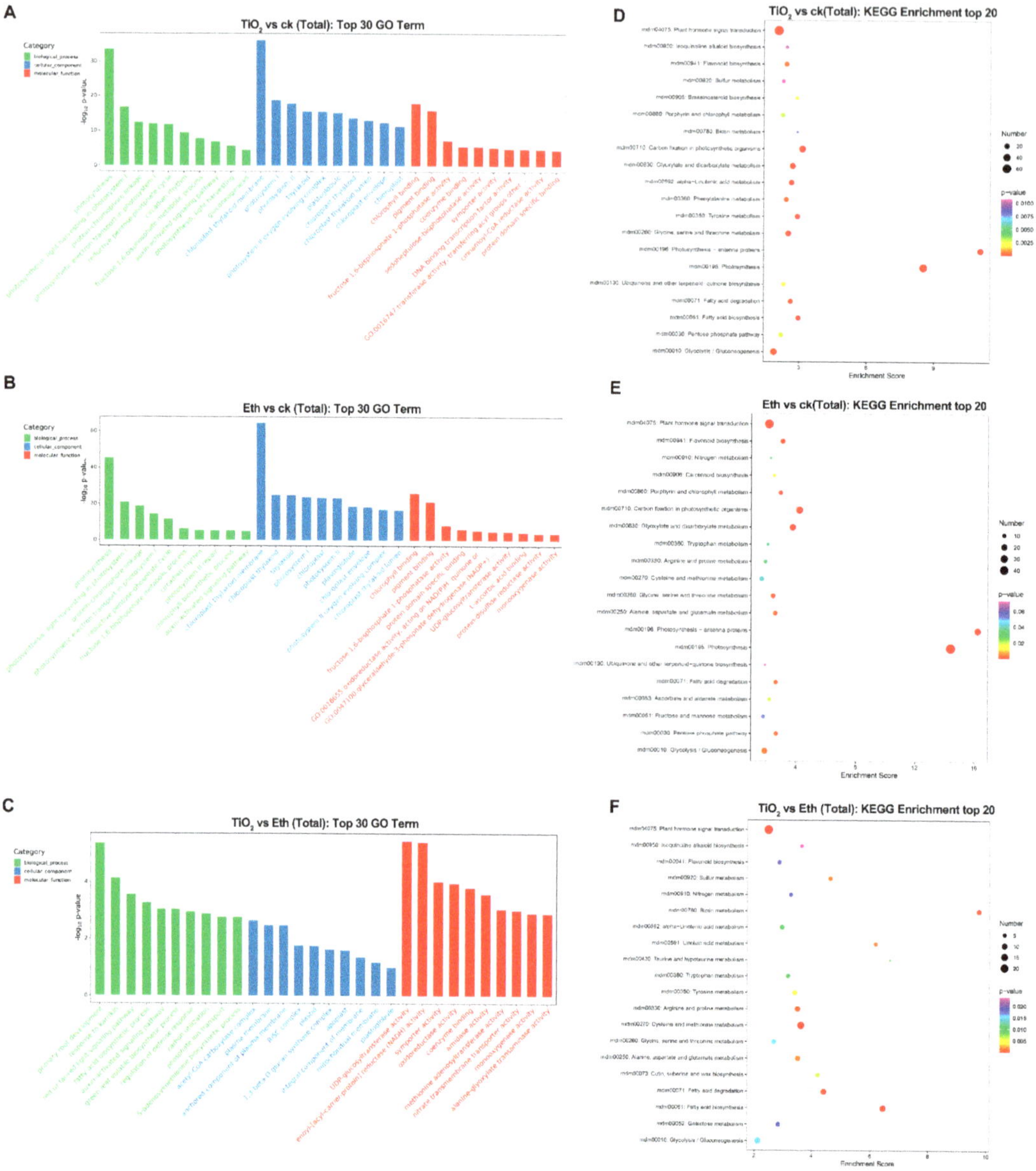

Figure 4. GO classification and KEGG pathway enrichment analyses of differentially expressed genes (DEGs) under the different treatments. Top 30 GO classification analysis of the genes identified under the biological process, molecular function, and cellular component categories between (**A**) NPs-TiO_2 vs. Control, (**B**) Ethylene vs. Control and (**C**) NPs-TiO_2 vs. Ethylene. Top 20 KEGG enrichment analysis of the DEGs between (**D**) NPs-TiO_2 vs. Control, (**E**) Ethylene vs. Control and (**F**) NPs-TiO_2 vs. Ethylene. Number of DEGs is represented by the size of the circle. ck represents Control, TiO_2 represents NPs-TiO_2, and Eth represents Ethylene.

2.5. *Differential Expression of Genes and Enzyme Activity Associated with Anthocyanin Biosynthesis in Exocarp*

Figure 5 shows the interaction expression of the major genes involved in anthocyanin metabolism in apple peel under different treatments, based on their corresponding positions in the apple anthocyanin biosynthesis pathway for 30 DEGs. The levels of gene expression in fruit treated with both treatments were significantly up-regulated compared to control.

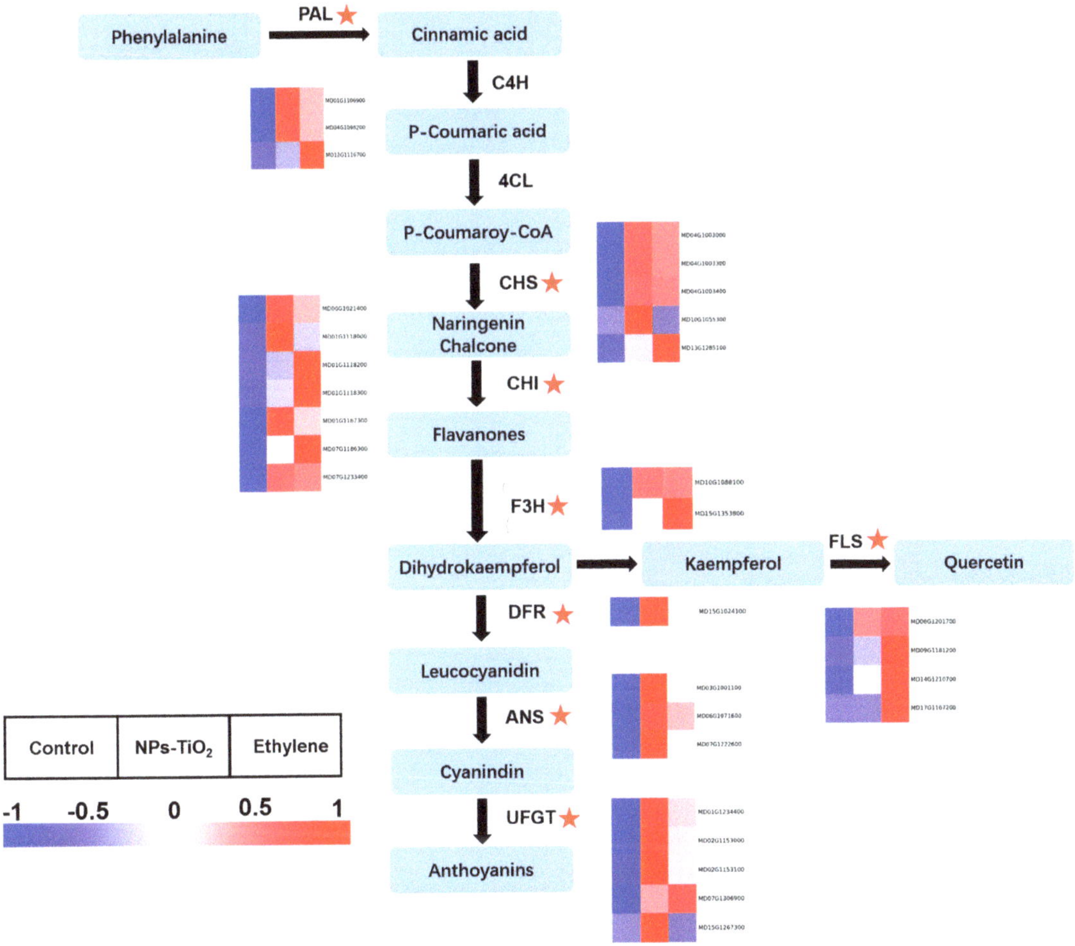

Figure 5. The regulation of flavonoids and anthocyanins biosynthesis pathways in apple under different treatments during light exposure. Heatmaps of DEGs in transcripts from flavonoids and anthocyanins biosynthesis pathway. PAL, phenylalanine ammonia-lyase; CHS, chalcone synthase; CHI, chalcone isomerase; F3H, flavanone 3 O-hydroxylase; DFR, dihydroflavonol 4 reductase; ANS, anthocyanidin synthase; UFGT, UDP glucose: flavonoid-3-*O*-glucosyltransferase; FLS, flavonoid 3′-monooxygenase. Red five-pointed stars represent significant DEGs in signaling pathways.

To verify the effects of the NPs-TiO_2 and ethylene treatments on the expression levels of genes involved in anthocyanin production in apple during light exposure, we examined the expression of eight genes that showed significant changes in expression over the treatment period (Figure 6). The relative expression levels of seven structural genes involved in the anthocyanin synthesis pathway, *MdPAL*, *MdCHS*, *MdCHI*, *MdF3H*, *MdDFR*, *MdUFGT*,

MdLDOX, and the key transcription factor *MdMYB1*, are shown in Figure 6A–H. The expression of *MdPAL*, *MdF3H*, *MdLDOX*, and *MdMYB1* first increased and then decreased in the NPs-TiO_2 and ethylene treatments. The expression of *MdCHS*, *MdCHI*, *MdDFR*, and *MdUFGT* continuously increased in the NPs-TiO_2 and ethylene treatment groups.

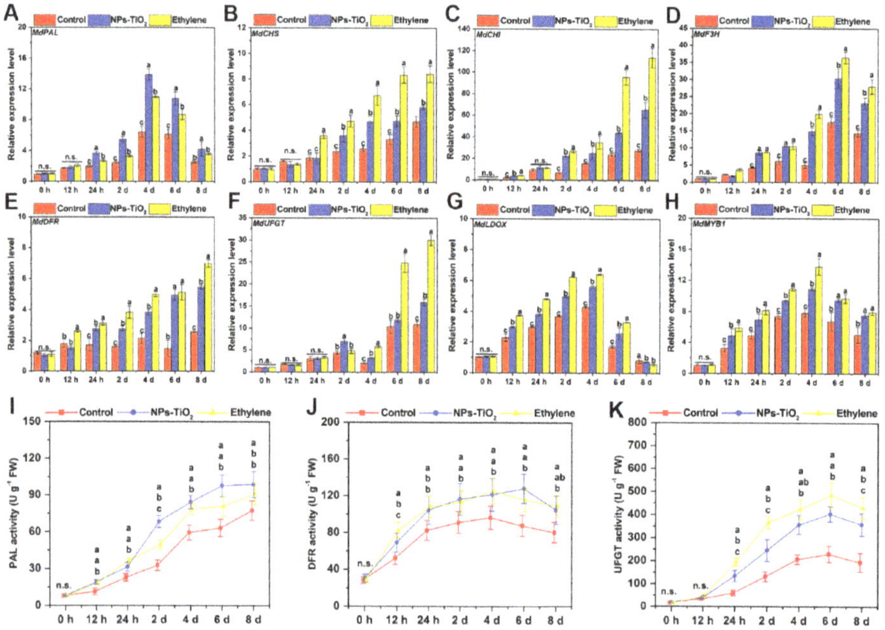

Figure 6. Expression of genes (**A–H**) and enzyme activities related to anthocyanin biosynthesis (**I–K**) in the 'Red Delicious' apple exocarp tissues during light exposure with NPs-TiO_2 and ethylene treatments. Error bars, mean ± standard error (n = 3). Different letters indicate significant differences among treatments in the same period using Duncan's test ($p < 0.05$), n.s. represents no significant difference ($p \geq 0.05$). PAL, phenylalanine ammonia-lyase; CHS, chalcone synthase; CHI, chalcone isomerase; F3H, flavanone-3-*O*-hydroxylase; DFR, dihydroflavonol 4-reductase; UFGT, UDP glucose: flavonoid-3-*O*-glucosyltransferase; LDOX, leucoanthocyanidin dioxygenase; MYB, R2R3-MYB transcription factor.

To clarify the effects of the NPs-TiO_2 and ethylene treatments on the activities of enzymes involved in anthocyanin synthesis, the activities of three enzymes involved in anthocyanin synthesis were measured. As shown in Figure 6I, significant differences ($p < 0.05$) in phenylalanine ammonia-lyase (PAL) activity were observed in peels treated with NPs-TiO_2 and ethylene in the early stage of the treatment period (0–24 h). The activity of PAL in the NPs-TiO_2 treatment increased rapidly at 2 d, and PAL activity was 2.16 times higher at 2 d than at 24 h. The PAL activity was also significantly higher ($p < 0.05$) in the NPs-TiO_2 treatment than in the ethylene treatment (1.25 times) and control (1.43 times). Changes in the activity of dihydroflavonol 4-reductase (DFR) in NPs-TiO_2 and ethylene treatments first increased and then decreased, and the activity of DFR was significantly ($p < 0.05$) higher in the two treatments than in the control at 12 h during the light exposure (Figure 6J). The activity of UDP-glucose: flavonoid-3-*O*-glucosyltransferase (UFGT) was higher in the ethylene treatment than in the NPs-TiO_2 treatment and the control at 24 h; UFGT activity peaked at day 6 and decreased at day 8 (Figure 6K).

2.6. Differential Expression of Genes Related to Ethylene and UV Light Signals Pathway

Both ethylene and UV light can promote apple pigment accumulation. We also analyzed the expression of genes involved in the ethylene pathway during light exposure (Figure 7A–D). The results showed that the expression of *MdACS1*, *MdACO1*, *MdERF3* and *MdEIL1* were markedly up-regulated in the ethylene-treated fruit compared to the other two groups. Meanwhile, there were no significant ($p \geq 0.05$) changes in the expression patterns of ethylene biosynthesis and signaling related genes under NPs-TiO_2 treatment. We further analyzed the expression levels of genes involved in UV-B signaling (Figure 7E–G). The expression levels of *MdCOP1*, *MdHY5*, and *MdUVR8* in the NPs-TiO_2 treatment were highest in the early and middle stages of the treatment period (24 h–4 days), and were markedly higher ($p < 0.05$) than those in the other two treatments.

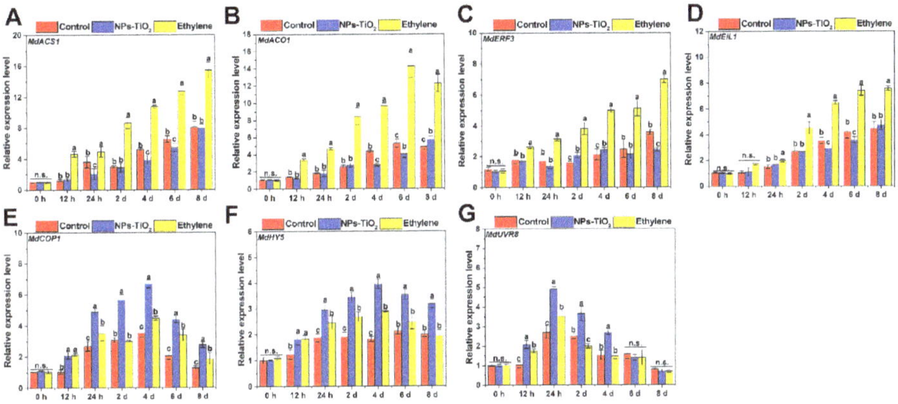

Figure 7. Expression of genes related to ethylene (**A–D**) and UV light signals pathway (**E–G**) in the 'Red Delicious' apple exocarp tissues during light exposure with the NPs-TiO_2 and ethylene treatments. Error bars, mean ± standard error (n = 3). Different letters indicate significant differences among treatments in the same period using Duncan's test ($p < 0.05$), n.s. represents no significant difference ($p \geq 0.05$). ACS1, 1-aminocyclopropane-1-carboxylate synthase 1; ACO1: 1-aminocyclopropane-1-carboxylate oxidase 1; EIL1: ethylene insensitive3-like 1; ERF3: ethylene response factors 3; COP1, constitutively photomorphogenic 1; HY5, elongated hypocotyl 5; UVR8, UV resistance locus 8.

2.7. Changes in ROS Levels and Membrane Lipid Peroxidation during Light Exposure

To clarify the effects of the NPs-TiO_2 and ethylene treatments on oxidative damage to the cell membrane and the degree of membrane lipid peroxidation in the fruit epidermis during light exposure, we measured the cell membrane permeability (Figure 8A) and malondialdehyde (MDA) content (Figure 8B) of the exocarp. At 12 h, the cell membrane permeability of the exocarp treated with the two treatments increased notably ($p < 0.05$), and the cell membrane permeability was 3.07% higher in the ethylene treatment than in the NPs-TiO_2 treatment. Treatment with NPs-TiO_2 resulted in a rapid increase in the cell membrane permeability of the peel on day 4, and the cell membrane permeability of the exocarp was significantly higher ($p < 0.05$) in the NPs-TiO_2 treatment than in the ethylene treatment and control (Figure 8A). The content of MDA increased during the light exposure, and the increase was most pronounced in the ethylene treatment, followed by the NPs-TiO_2 treatment and the control (Figure 8B); at the end of the treatment period, the content of MDA was markedly higher ($p < 0.05$) in both treatments than in the control.

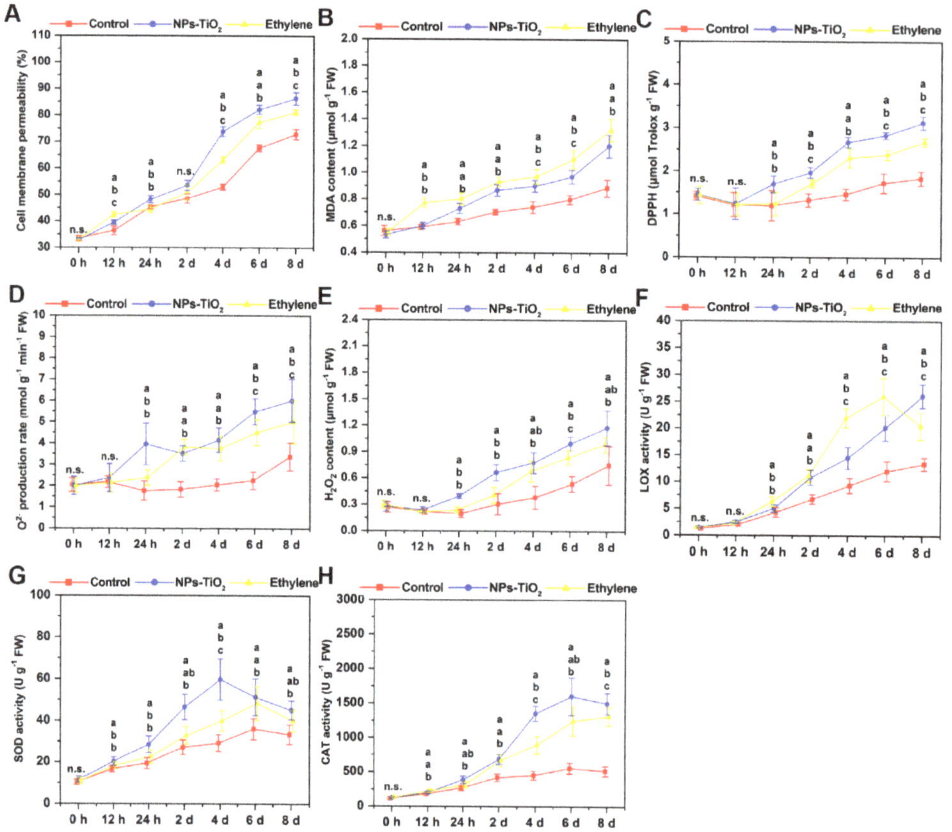

Figure 8. Effects of two treatments on (**A**) Cell membrane permeability, (**B**) MDA content, (**C**) DPPH, (**D**) O_2^- production rate and (**E**) H_2O_2 content, (**F**) LOX activity, (**G**) SOD activity and (**H**) CAT activity during light cultivation of the 'Red Delicious' apples. Error bars, mean ± standard error ($n = 3$). Different letters indicate significant differences among treatments in the same period using Duncan's test ($p < 0.05$), n.s. represents no significant difference ($p \geq 0.05$).

The total antioxidant capacity of the exocarp tissue during light exposure was determined via a 2,2-diphenyl-1-picryl-hydrazyl-hydrate (DPPH) assay. The content of DPPH in the peel decreased slightly at 12 h and then increased over the treatment period (Figure 8C). The relative content of DPPH was 2.17 times higher in the NPs-TiO_2 treatment on the eighth day than at the start of the treatment period; the content of DPPH was significantly higher ($p < 0.05$) in the NPs-TiO_2 treatment than in the ethylene treatment (1.87 times) and control (1.30 times). Changes in the O_2^- production rate and the H_2O_2 content were similar over the treatment period (Figure 8D,E). At 24 h, the H_2O_2 content was significantly higher in the NPs-TiO_2 treatment ($p < 0.05$) than in the ethylene treatment and control, and the H_2O_2 content increased continuously after 24 h.

The activity of lipoxygenase (LOX) increased in all fruit after 24 h of treatment; LOX activity was significantly higher ($p < 0.05$) in the ethylene treatment than in the NPs-TiO_2 treatment and control, and LOX activity peaked in the ethylene treatment on the sixth day at 26.04 ± 3.42 U g^{-1}. The activity of LOX on the eighth day was 1.27 and 1.95 times ($p < 0.05$) higher in the NPs-TiO_2 treatment than in the ethylene treatment and control, respectively (Figure 8F). Changes in the activity of catalase (CAT) and superoxide dismutase (SOD) in fruit in the NPs-TiO_2 treatment were similar over the treatment period, and CAT and SOD activity was higher in the NPs-TiO_2 treatment than in the ethylene treatment and the control (Figure 8G,H). In the NPs-TiO_2 treatment group, the activity of SOD peaked (59.80 ± 9.97 U g^{-1}) on the fourth day, and the activity of CAT peaked (1603.91 ± 272.66 U g^{-1}) on the sixth day.

3. Discussion

The appearance quality of fruit has a major effect on its economic value, and dysplasia and instability of the color of fruit during cultivation and production have negative economic effects. NPs-TiO_2 are semiconductor materials with a band gap of approximately 3.2 eV, corresponding to a wavelength of approximately 390 nm, and the crystal surface of the anatase phase is the most photocatalytically active [33]. Therefore, UV absorption by the apple peel is enhanced after application with NPs-TiO_2 nano-colloids resulting in higher UV effective absorption. This consequently leads to the accumulation of more pigments. In this study, we found that the effect of the NPs-TiO_2 treatment on fruit quality was weaker than that of the ethylene treatment. NPs-TiO_2 treatment had less impact on fruit quality during the accelerated pigmentation process as reflected by the insignificant ($p \geq 0.05$) reduction in fruit firmness, and the insignificant increase in TSS and the solid–acid ratio. In addition, the ethylene release rate and fruit respiration rate also indicated that the effects of NPs-TiO_2 treatment alleviated the post-ripening process during the post-harvest treatment period and was weaker compared with the effects of the ethylene spray treatment. These results indicate that NPs-TiO_2 promoted peel color and had a weaker effect on fruit quality than the ethylene treatment. Light plays an important role in fruit development. Light can affect physiological changes in fruits, especially the synthesis of phenolic substances in the later ripening stage [34]. Primary phenolics in apples are secondary metabolites with antioxidant properties, and the total phenol content and total flavonoid content in some apple varieties are significantly positively correlated with levels of antioxidant activity [35]. Anthocyanins and proanthocyanidins are produced through the flavonoid pathway and are important secondary metabolites that control apple appearance quality, and flavonoids are components of the phenylpropane system [36,37]. Interactions between many important enzymes in the flavonoid pathway are assumed to be the cause of anthocyanin accumulation [38]. In our study, changes in the content of anthocyanins and flavonoids of fruit were more gradual in the NPs-TiO_2 treatment than in the ethylene treatment, and decreases in the content of proanthocyanidins in the ethylene and NPs-TiO_2 treatments were similar during the two treatments. This suggests that the activity of the enzymes and transcription factors involved in the flavonoid synthesis pathway were higher under ethylene treatment than in the NPs-TiO_2 treatment.

Light induces anthocyanin biosynthesis by promoting the expression of key enzyme genes involved in the anthocyanin biosynthesis pathway in apples, and the induction of UV light is extremely important [4,7,8,39]. In this study, KEGG analysis revealed that the DEGs in the two treatments were primarily enriched in 'Flavonoid synthesis' pathways, which also corresponded to changes in the content of anthocyanins and flavonoids over the treatment period. This is consistent with the results of previous studies of apple fruits that were treated with UV light, which promoted the accumulation of anthocyanins in the flavonoid metabolic pathway [40,41].

Anthocyanins are derived from the phenylpropanol biosynthesis pathway, and several enzymes are involved in anthocyanin biosynthesis; the functions of anthocyanins have been elucidated [42]. These structural genes are regulated by several transcription factors at the transcriptional and translational levels under different environmental conditions [4]. The proteins encoded by MYB family genes are components of the MYB-bHLH-WD40/WDR protein complex, and MdMYB1 plays an essential role in this process [43–45]. We examined seven structural genes involved in anthocyanin production and, with the exception of *MdPAL*, the expression of these genes was higher in the ethylene treatment than in the NP-TiO_2 treatment. In addition, the transcript levels of the MYB, bHLH, and WD40 group genes were also typically quite different under the two treatments compared to the control. The activity of the three anthocyanin-related enzymes PAL, DFR, and UFGT was determined, and changes in their activity were consistent during the treatment period. Ethylene biosynthesis begins with the synthesis of ACC by 1-aminocyclopropane-1-carboxylate synthase (ACS) from S-adenosyl methionine, which is then oxidized to ethylene by ACC oxidase (ACO). The *ACS* and *ACO* genes have been identified as important factors in the ethylene production in fruit and in the ripening of fruit [46,47]. Ethylene is an important plant hormone, and numerous studies have shown that it can promote anthocyanin accumulation in apple [44,48–50]. In apples, *MdEIL1* promotes anthocyanin biosynthesis by directly binding to the *MdMYB1* promoter, and MdERF3 interacts with MdMYB1 to positively regulate ethylene production in the fruits [44]. The expression of genes related to ethylene biosynthesis was remarkably higher ($p < 0.05$) in the ethylene treatment (0.1% ethephon, w/v) than in the NPs-TiO_2 treatment and the control. Correspondingly, the ethylene productions were higher in the ethylene treatment, and RNA-Seq data also indicated that the expression levels of structural genes during ethylene biosynthesis were more active in ethylene-treated fruit (Supplementary Table S3). All the abovementioned findings showed that ethylene treatment promoted anthocyanin accumulation mainly dependent on the ethylene pathway. In addition, we found that there were no significant changes in the expression patterns of ethylene production and signaling-related genes under NPs-TiO_2 treatment. UV light can promote apple anthocyanin accumulation [7,8]. Meanwhile, photocatalytic nanomaterials promote fruit coloration by improving the absorption of UV light [30]. Therefore, we further analyzed the expression levels of relevant gene expression in the light signaling pathway. In apple, MdCOP1 and MdHY5 encode key transcription factors involved in anthocyanin biosynthesis and plant photosynthesis [51,52]. In addition, MdUVR8 is a UV-B photoreceptor that can interact with MdCOP1 and plays an essential role in photomorphogenesis [53]. Interestingly, we found the expression levels of these three genes were higher in fruit treated with NPs-TiO_2 than in fruit treated with ethylene. These findings further indicate that anthocyanin biosynthesis in fruit is induced by UV light signals under NPs-TiO_2 treatment, which differ from those induced by ethylene.

ROS are signaling molecules that are crucial for controlling plant development, senescence, and growth. Recent investigations have shown that ROS may play a role in controlling the production of anthocyanins in apples [54,55]. According to previous studies, UV radiation stimulates the production of ROS by regulating the activity of NADPH oxidase in the cytoplasmic membrane of the exocarp, thereby enhancing anthocyanin biosynthesis [56,57]. Elevated levels of ROS can induce lipid peroxidation of the cell membrane, which is mainly catalyzed by LOX, and MDA is one of the end products of lipid peroxidation of the cell membrane. Intracellular ROS levels and cell membrane permeability were higher in the NPs-TiO_2 treatment than in the ethylene treatment and the control during the treatment period. However, the MDA content was slightly higher in the ethylene treatment than in the NPs-TiO_2 treatment during the treatment period; a sharp increase in LOX enzyme activity was observed at 4 days and 6 days. We evaluated levels of antioxidant activity in fruit during the treatment period and found that the levels of antioxidant enzymes and the DPPH content were higher in the NPs-TiO_2 treatment than in the ethylene treatment and the control during the treatment period. This also explains why the MDA content was lower in the NPs-TiO_2 treatment group than in the ethylene treatment group. In sum, treatment with NPs-TiO_2 and ethylene induced the accumulation of anthocyanins in fruit during light exposure. The antioxidant capacity and ROS levels of fruits were higher and levels of membrane lipid peroxidation were lower in the NPs-TiO_2 treatment than in the ethylene treatment. Additional research is needed to determine whether the ROS generated via the photocatalytic activity of NPs-TiO_2 or the absorption and use of UV light by NPs-TiO_2 are the principal drivers of anthocyanin biosynthesis in fruit.

4. Materials and Methods

4.1. Preparation and Characterization of NPs-TiO_2

NPs-TiO_2 were thermally synthesized following the methods described in previous studies [30,58]. The prepared NPs-TiO_2 were characterized by scanning electron microscopy (SEM), transmission electron microscopy (TEM), dynamic light scattering (DLS) and X-ray crystal diffraction (XRD). The obtained NPs-TiO_2 had an irregular spherical shape and a crystal conformation with a particle size of 10.34 $\pm$ 3.34 nm (Figure S1). After obtaining dry NPs-TiO_2 powder, NPs-TiO_2 colloidal solution was prepared via an ultrasonic process [30].

4.2. Fruit Material and Experimental Treatments

Apple fruit (*Malus domestica* Borkh. 'Red Delicious') was collected from an orchard in Tai'an, Shandong Province, China. The fruits were bagged at the young fruit stage (40 days after the flowering), and the orchard was managed using standard practices during growth and development. The fruits used in this study were harvested at 150 days after flowering (DAF). Apples of the same size and with no obvious signs of physiological disease were collected and stored in soft-bottom foam boxes. They were then transported to the laboratory of Shandong Agricultural University the same day that they were collected for subsequent experimental treatments. The apples were divided into three groups, with 30 apples in each group, and transferred into bags in low-light environments. Following previous studies [44,59], 0.1% (w/v) ethephon solution and 0.1% (w/v) NPs-TiO_2 colloids were sprayed on apples; in the control treatment, the same amount of deionized water was applied. After the surface water dried naturally, the cells were placed into three independent light incubators (25 °C, 70 μmol m^{-2} s^{-1} intensity of illumination. and 330 $\pm$ 20 nm wavelength) for light cultivation for 8 days. The peel tissue was sampled at 0 h, 12 h, 24 h, 2 days, 4 days, 6 days and 8 days. Three fruit skin samples were taken from each treatment group at each time period and discarded immediately after the skin samples were taken. The samples were rapidly frozen with liquid nitrogen and stored at −80 °C in a refrigerator for subsequent experiments.

4.3. Determination of Fruit Physiological Quality

The exocarp color of the apple was determined using a chroma meter at the indicated times (0 h,12 h, 24 h, 2 days, 4 days, 6 days and 8 days). Fruit treated with light for 8 days were measured for firmness, total soluble solids (TSS), total titratable acid (TA) and solid–acids ratio physiological qualities as described in the previous study [59]. The respiration rate and ethylene production of the experiment fruit at the indicated times (0 h, 4 days, 8 days) were measured using a CO_2 detector (Testo535-CO_2, Testo Inc., Shanghai, China) and a gas chromatograph (GC-2014, Shimadzu, Toyko, Japan), respectively, as described in previous studies [60]. A total of six biological replicates were performed for each treatment in the above experiments.

4.4. Determination of Secondary Metabolites

The anthocyanin content in exocarp was determined using the hydrochloric acid–methanol method with reference to the previous study [61]. The supernatant's absorbance was calculated at 530, 620, and 650 nm. Anthocyanin content was expressed as 0.1 change in optical density (μg g^{-1} FW).

The content of proanthocyanidins was determined according to the method of previous research with slight modification [62]. At a wavelength of 640 nm, the absorbance readings of reaction system were recorded. The standard curve was made of catechin (μg g^{-1} FW).

The flavonoids and the total phenolics content were determined with reference to previous studies [63]. For the flavonoids content, the absorbance value of reaction solution at 430 nm wavelength was recorded. The rutin-based standard curve was used (μg g^{-1} FW). For the total phenolics content, the absorbance value of reaction solution at 725 nm wavelength was recorded. The gallic-acid-based standard curve was used (μg g^{-1} FW).

4.5. RNA-Seq Analysis

RNA was extracted from exocarp tissue exposed to 8 days of light, and purified RNA was collected in triplicate for each treatment group. OE Biotech Co., Ltd. carried out the sequencing and analysis of the transcriptome (Shanghai, China). Trimmomatic was used to process the raw data [64]. To achieve the clean reads, the low-quality reads and reads containing ploy-N were eliminated. A total of 60.47 G of clean data was obtained. Effective data for each sample ranged from 6.57G to 6.96 G, with Q30 bases ranging from 91.2% to 91.98%, and GC fractions ranging from 44.45% to 47.62% (Supplementary Table S1). Following that, hisat2 was used to map the clean reads to the reference genome [65]. By using the approaches from previous research [66–70], quantitative analysis of gene expression, differentially expressed genes analyses (DEGs), Gene Ontology analyses (GO) and Kyoto Encyclopedia of Genes and Genomes (KEGG) enrichment analyses were carried out. The threshold for significantly altered expression was chosen at $p < 0.05$ and foldchange > 2 or foldchange < 0.5. The RNA-seq raw data were submitted to NCBI with the following ID number: PRJNA947803.

4.6. Real-Time qPCR Expression Analysis

RNA plant plus Reagent kit (Tiangen, Beijing, China) was used to extract RNA from apple exocarp tissue according to the corresponding experimental methods. After obtaining the RNA of the pure sample, the PrimeScript cDNA synthesis kit (Tiangen, Beijing, China) was used for subsequent reverse transcription experiments to obtain the cDNA template of the sample. The cDNA template was subsequently diluted to a concentration of approximately 2 ng μL^{-1}. The qRT-PCR was performed using the $2^{-\Delta\Delta Ct}$ calculation method, and the results were normalized using the *MdActin* gene of apple as the reference gene. The qRT-PCR procedures were set according to the experimental methods of previous studies [52]. The primer sequences for all genes are provided in Supplementary Table S2.

4.7. Determination of Malonaldehyde (MDA) Content, Cell Membrane Permeability and Reactive Oxygen Species (ROS) Levels

The relative cell membrane permeability of the exocarp tissue was detected using an electrical conductivity meter (DDS-307A, INESA Scientific Inc., Shanghai, China) at the indicated times (0 h, 12 h, 24 h, 2 days, 4 days, 6 days and 8 days). The relative cell membrane permeability is expressed as % [60].

The content of MDA in exocarp was determined by trichloroacetic acid (TBA) method with slight modification according to the previous study [71]. The sample's absorbance value at wavelengths of 600 nm, 532 nm, and 450 nm was calculated (μmol g^{-1} FW).

The production rate of superoxide anion and the content of hydrogen peroxide (H_2O_2) in exocarp were determined by Superoxide Anion kit and Hydrogen Peroxide detection kit (Suzhou Keming Biotechnology Co., Ltd., Suzhou, China), respectively. The absorption value of the reaction system at a wavelength of 530 nm was recorded by a microplate reader to determine the superoxide anion production rate. One hour passed without any static reaction occurring in the centrifuge tubes of the control group. The result is denoted as nmol g^{-1} min^{-1} FW. For H_2O_2 content, the absorbance value of reaction solution at 415 nm wavelength was recorded (μmol g^{-1} FW).

4.8. Determination of Total Antioxidant Capacity

The scavenging activity of 2,2-diphenyl-1-picryl hydrazide (DPPH) free radical in exocarp tissues was measured by using the previous method [72]. The same mass fraction of ethanol solution was used to generate the standard solution curve of Trolox. At a wavelength of 515 nm, the reaction solution's absorbance was recorded. The results of DPPH relative content are expressed as μmol Trolox g^{-1} FW.

4.9. Determination of Enzyme Activity

For LOX, one gram of frozen materials was crushed in 10 mL of 0.05 mM of phosphatic buffer solution (PBS, PH 7.8, containing 50 mM β-mercaptoethanol), and the homogenate was centrifuged for 10 min (4 °C, 12,000 rpm). The substrate for the enzymatic reaction was prepared by adding 0.5 mL linoleic acid, 0.25 mL Tween 20, and 1 mL NaOH (1 M) to 20 mL borate buffer, followed by adjusting the pH to 9.0 and diluting to 500 mL. The reaction system was prepared by mixing 0.2 mL of the supernatants with 0.8 mL of the enzyme substrates [73]. The absorbance value of reaction system at 234 nm wavelength was detected using a microplate reader (U g^{-1} FW)

The extraction and determination methods of superoxide dismutase (SOD) and catalase (CAT) were referred to previous studies [63]. The ability of the samples to inhibit the photo-chemical reduction of Nitro-Blue Tetrazolium (NBT) was analyzed to assay the enzyme activity of SOD. The absorbance value of reaction solution at 560 nm wavelength was detected by microplate reader (U g^{-1} FW). For CAT, the absorbance value of reaction solution at 240 nm wavelength was detected by microplate reader. A unit of enzyme activity was defined as 1 nmol H_2O_2 degradation per minute per gram of tissue catalyzed (U g^{-1} FW).

The enzyme activities of l-Phenylalanine Ammonia-Lyase (PAL), Dihydroflavonol 4-Reductase (DFR) and UDP-glycose flavonoid glycosyltransferase (UFGT) enzyme activity were determined by enzyme activity detection kit (Hefei Laier Biotechnology, Inc., China). For PAL, the absorbance of reaction solution at a wavelength of 290 nm was measured using a microplate reader. Each gram of tissue in each milliliter reaction system changes the absorbance value at 290 nm by 0.1 per minute, which is one unit of enzyme activity (U g^{-1} FW). For DFR, the absorbance value of reaction solution at a wavelength of 500 nm was measured using a microplate reader. A unit of enzyme activity was defined as 1 mmol catechin per gram of tissue per minute of tissue catalyzed (U g^{-1} FW). For UFGT, the absorbance value of reaction solution at 340 nm wavelength was detected by microplate reader (U g^{-1} FW). A unit of enzyme activity was defined as 1 nmol NADH per minute per gram of tissue catalyzed.

4.10. Statistical Analysis

The experiment was conducted in a completely random manner. The data were evaluated by One-way analysis of variance (ANOVA) using Excel 2019 and Origin 2021 software. At least three biological replicates were used in each experiment. All data were reported as mean ± standard error (SE). The ANOVA was performed by $p < 0.05$. The difference was significant by least partial square test (LSD).

5. Conclusions

We analyzed the effects of treatment with NPs-TiO_2 and ethylene on the anthocyanin biosynthesis and ROS metabolism in ripe 'Red Delicious' apples under light exposure. The application of both treatments significantly promoted the deposition of pigments in the peel during light exposure, and NPs-TiO_2 had a minor effect on the quality of the fruit during the treatment period. Furthermore, ethylene treatment promoted fruit coloring mainly by up-regulating the expression of ethylene-related genes, whereas NPs-TiO_2 promoted fruit coloring mainly by up-regulating the expression of UV-radiation-related genes. The ROS levels and antioxidant capacity of the treated fruit were higher in the NPs-TiO_2 treatment group than in the ethylene treatment group during the treatment period, and the degree of membrane lipid peroxidation was lower in fruit in the NPs-TiO_2 treatment than in the ethylene treatment. These results imply the potential utility of NPs-TiO_2 in enhancing apple coloration. Moreover, the transcriptomic data offer novel insights into the molecular mechanisms underlying the promotion of apple coloration via NPs-TiO_2 treatment.

Supplementary Materials: The following supporting information can be downloaded at: https://www.mdpi.com/article/10.3390/foods12163137/s1, Figure S1: NPs-TiO_2 characterization. (A) and (B) SEM. (C) TEM. (D) Particle size distribution. (E) X-RD; Figure S2: Effects of NPs-TiO_2 and ethylene treatments on (A) respiratory rate and (B) ethylene production during light exposure in apples. Error bars, mean ± standard error (n = 3). Different letters indicate significant differences among treatments in the same period using Duncan's test ($p < 0.05$), n.s. represents no significant difference ($p \geq 0.05$); Figure S3: qRT-PCR validation of DEGs in the peels of different treatments detected by RNA-Seq. Error bars, mean ± standard error (n = 3). Different letters indicate significant differences among treatments in the same period using Duncan's test ($p < 0.05$); Figure S4: Heatmap representation of the expression patterns of genes associated with the MYB-bHLH-WD40 protein complex. Rows and columns in the heat map represent genes and samples collected. The color bar represents the FPKM average values. MYB, MYB family members; bHLH, basic helix–loop–helix; WD40, WD40-repeat proteins; Table S1: Primers for qRT-PCR; Table S2: Quality evaluation of RNA-Seq data; Table S3: DEGs related to ethylene biosynthesis pathway in apple.

Author Contributions: Y.W.: Conceptualization, Methodology, Investigation, Data curation, Writing—original draft. G.C., D.W., J.Z., C.Y. and X.W.: Validation, Writing—review and editing, Supervision. H.L.: Supervision, Writing—review and editing. All authors have read and agreed to the published version of the manuscript.

Funding: This study was supported by the National Natural Science Foundation of China (32172538), the major basic research projects of Shandong Province (ZR2020ZD43), the China Agriculture Research System of MOF and MARA (CARS-27), the Shandong Province Key R&D Program (2021CXGC010802), and Taishan scholar project (LJNY202026).

Data Availability Statement: The data used to support the findings of this study can be made available by the corresponding author upon request.

Conflicts of Interest: The authors declare no conflict of interest.

References

1. Muder, A.; Garming, H.; Dreisiebner-Lanz, S.; Kerngast, K.; Rosner, F.; Kličková, K.; Kurthy, G.; Cimer, K.; Bertazzoli, A.; Altamura, V.; et al. Apple production and apple value chains in Europe. *Eur. J. Hortic. Sci.* **2022**, *87*, 1611–4434. [CrossRef]
2. FAO. FAOSTAT Database 2021. 2021. Available online: http://www.fao.org/faostat/en/#data/QC (accessed on 7 July 2023).
3. Hamadziripi, E.; Muller, M.; Theron, K.I.; Steyn, W.J. Consumer preference for apple eating quality and taste in relation to canopy position. *Acta Hortic.* **2017**, *1058*, 253–260. [CrossRef]
4. An, J.P.; Wang, X.F.; Hao, Y.J. BTB/TAZ protein MdBT2 integrates multiple hormonal and environmental signals to regulate anthocyanin biosynthesis in apple. *J. Integr. Plant Biol.* **2020**, *62*, 1643–1646. [CrossRef] [PubMed]
5. Weber, S.; Damerow, L.; Kunz, A.; Blanke, M. Anthocyanin synthesis and light utilisation can be enhanced by reflective mulch-Visualisation of light penetration into a tree canopy. *J. Plant Physiol.* **2019**, *233*, 52–57. [CrossRef] [PubMed]
6. Funke, K.; Blanke, M. Spatial and Temporal Enhancement of Colour Development in Apples Subjected to Reflective Material in the Southern Hemisphere. *Horticulturae* **2021**, *7*, 2. [CrossRef]
7. Henry-Kirk, R.A.; Plunkett, B.; Hall, M.; McGhie, T.; Allan, A.C.; Wargent, J.J.; Espley, R.V. Solar UV light regulates flavonoid metabolism in apple (Malus x domestica). *Plant Cell Environ.* **2018**, *41*, 675–688. [CrossRef]
8. Fang, H.; Dong, Y.; Yue, X.; Hu, J.; Jiang, S.; Xu, H.; Wang, Y.; Su, M.; Zhang, J.; Zhang, Z.; et al. The B-box zinc finger protein MdBBX20 integrates anthocyanin accumulation in response to ultraviolet radiation and low temperature. *Plant Cell Environ.* **2019**, *42*, 2090–2104. [CrossRef]
9. Wang, Z.; Dilley, D.R. Aminoethoxyvinylglycine, combined with ethephon, can enhance red color development without over-ripening apples. *HortScience* **2001**, *36*, 328–331. [CrossRef]
10. Layne, D.R.; Jiang, Z.; Rushing, J.W. The Influence of Reflective Film and ReTain on Red Skin Coloration and Maturity of 'Gala' Apples. *Horttechnology* **2002**, *12*, 640–645. [CrossRef]
11. Iglesias, I.; Salvia, J.; Torguet, L.; Montserrat, R. The evaporative cooling effects of overtree microsprinkler irrigation on 'Mondial Gala' apples. *Sci. Hortic.* **2005**, *103*, 267–287. [CrossRef]
12. Fan, X.; Mattheis, J.P. Bagging 'Fuji' apples during fruit development affects color development and storage quality. *Hortscience* **1998**, *33*, 1235–1238. [CrossRef]
13. Awad, M.A.; Jager, A.d. Formation of flavonoids, especially anthocyanin and chlorogenic acid in 'Jonagold' apple skin: Influences of growth regulators and fruit maturity. *Sci. Hortic.* **2002**, *93*, 257–266. [CrossRef]
14. Saure, M.C. External control of anthocyanin formation in apple. *Sci. Hortic.* **1990**, *42*, 181–218. [CrossRef]
15. Whale, S.K.; Singh, Z. Endogenous Ethylene and Color Development in the Skin of 'Pink Lady' Apple. *J. Am. Soc. Hortic. Sci.* **2007**, *132*, 20–28. [CrossRef]

16. Forlani, S.; Masiero, S.; Mizzotti, C. Fruit ripening: The role of hormones, cell wall modifications, and their relationship with pathogens. *J. Exp. Bot.* **2019**, *70*, 2993–3006. [CrossRef] [PubMed]
17. Chen, M.; Gu, H.; Wang, L.; Shao, Y.; Li, R.; Li, W. Exogenous Ethylene Promotes Peel Color Transformation by Regulating the Degradation of Chlorophyll and Synthesis of Anthocyanin in Postharvest Mango Fruit. *Front. Nutr.* **2022**, *9*, 911542. [CrossRef]
18. Ali, A.; Yeoh, W.K.; Forney, C.; Siddiqui, M.W. Advances in postharvest technologies to extend the storage life of minimally processed fruits and vegetables. *Crit. Rev. Food Sci. Nutr.* **2018**, *58*, 2632–2649. [CrossRef]
19. Lowry, G.V.; Avellan, A.; Gilbertson, L.M. Opportunities and challenges for nanotechnology in the agri-tech revolution. *Nat. Nanotechnol.* **2019**, *14*, 517–522. [CrossRef]
20. Gogos, A.; Knauer, K.; Bucheli, T.D. Nanomaterials in plant protection and fertilization: Current state, foreseen applications, and research priorities. *J. Agric. Food Chem.* **2012**, *60*, 9781–9792. [CrossRef]
21. Verma, S.K.; Das, A.K.; Patel, M.K.; Shah, A.; Kumar, V.; Gantait, S. Engineered nanomaterials for plant growth and development: A perspective analysis. *Sci. Total Environ.* **2018**, *630*, 1413–1435. [CrossRef]
22. Schneider, J.; Matsuoka, M.; Takeuchi, M.; Zhang, J.; Horiuchi, Y.; Anpo, M.; Bahnemann, D.W. Understanding TiO_2 photocatalysis: Mechanisms and materials. *Chem. Rev.* **2014**, *114*, 9919–9986. [CrossRef]
23. Chen, H.; Nanayakkara, C.E.; Grassian, V.H. Titanium dioxide photocatalysis in atmospheric chemistry. *Chem. Rev.* **2012**, *112*, 5919–5948. [CrossRef] [PubMed]
24. Fujishima, A.; Zhang, X.; Tryk, D.A. TiO_2 photocatalysis and related surface phenomena. *Surf. Sci. Rep.* **2008**, *63*, 515–582. [CrossRef]
25. Pelaez, M.; Nolan, N.T.; Pillai, S.C.; Seery, M.K.; Falaras, P.; Kontos, A.G.; Dunlop, P.S.M.; Hamilton, J.W.J.; Byrne, J.A.; O'Shea, K.; et al. A review on the visible light active titanium dioxide photocatalysts for environmental applications. *Appl. Catal. B Environ.* **2012**, *125*, 331–349. [CrossRef]
26. Dalrymple, O.K.; Stefanakos, E.; Trotz, M.A.; Goswami, D.Y. A review of the mechanisms and modeling of photocatalytic disinfection. *Appl. Catal. B Environ.* **2010**, *98*, 27–38. [CrossRef]
27. López-Moreno, M.L.; Cedeño-Mattei, Y.; Bailón-Ruiz, S.J.; Vazquez-Nuñez, E.; Hernandez-Viezcas, J.A.; Perales-Pérez, O.J.; De la Rosa, G.; Peralta-Videa, J.R.; Gardea-Torresdey, J.L. Environmental behavior of coated NMs: Physicochemical aspects and plant interactions. *J. Hazard. Mater.* **2018**, *347*, 196–217. [CrossRef]
28. Uyguner Demirel, C.S.; Birben, N.C.; Bekbolet, M. A comprehensive review on the use of second generation TiO_2 photocatalysts: Microorganism inactivation. *Chemosphere* **2018**, *211*, 420–448. [CrossRef]
29. Xie, J.; Hung, Y.-C. UV-A activated TiO_2 embedded biodegradable polymer film for antimicrobial food packaging application. *LWT-Food Sci. Technol.* **2018**, *96*, 307–314. [CrossRef]
30. Wang, C.; Wang, Y.; Sun, Q.; Hao, Y. Nanoscale UV Absorber Boosting Coloration of Apple Skin. *ACS Sustain. Chem. Eng.* **2019**, *7*, 16295–16300. [CrossRef]
31. Grange, M.L.; Wand, S.J.E.; Theron, K.I. Effect of Kaolin Applications on Apple Fruit Quality and Gas Exchange of Apple Leaves. *Acta Hortic.* **2004**, *636*, 545–550. [CrossRef]
32. Takos, A.M.; Jaffé, F.W.; Jacob, S.R.; Bogs, J.; Robinson, S.P.; Walker, A.R. Light-induced expression of a MYB gene regulates anthocyanin biosynthesis in red apples. *Plant Physiol.* **2006**, *142*, 1216–1232. [CrossRef] [PubMed]
33. Fang, W.Q.; Gong, X.Q.; Yang, H.G. On the Unusual Properties of Anatase TiO_2 Exposed by Highly Reactive Facets. *J. Phys. Chem. Lett.* **2011**, *2*, 725–734. [CrossRef]
34. Sun, S.; Xin, L.; Gao, H.; Wang, J.; Li, P. Response of phenolic compounds in 'Golden Delicious' and 'Red Delicious' apples peel to fruit bagging and subsequent sunlight re-exposure. *Sci. Hortic.* **2014**, *168*, 161–167. [CrossRef]
35. Vieira, F.; Borges, G.; Copetti, C.; Pietro, P.; Nunes, E.; Fett, R. Phenolic compounds and antioxidant activity of the apple flesh and peel of eleven cultivars grown in Brazil. *Sci. Hortic.* **2011**, *128*, 261–266. [CrossRef]
36. Winkel-Shirley, B. lavonoid Biosynthesis. A Colorful Model for Genetics, Biochemistry, Cell Biology, and Biotechnology. *Plant Physiol.* **2001**, *126*, 485–493. [CrossRef] [PubMed]
37. Grotewold, E. *The Science of Flavonoids*; Springer: New York, NY, USA, 2006.
38. Ju, Z. Fruit bagging, a useful method for studying anthocyanin synthesis and gene expression in apples. *Sci. Hortic.* **1998**, *77*, 155–164. [CrossRef]
39. An, J.P.; Wang, X.F.; Zhang, X.W.; Bi, S.Q.; You, C.X.; Hao, Y.J. MdBBX22 regulates UV-B-induced anthocyanin biosynthesis through regulating the function of MdHY5 and is targeted by MdBT2 for 26S proteasome-mediated degradation. *Plant Biotechnol. J.* **2019**, *17*, 2231–2233. [CrossRef] [PubMed]
40. Ding, R.; Che, X.; Shen, Z.; Zhang, Y. Metabolome and transcriptome profiling provide insights into green apple peel reveals light- and UV-B-responsive pathway in anthocyanins accumulation. *BMC Plant Biol.* **2021**, *21*, 351. [CrossRef]
41. Ryu, J.A.; Duan, S.; Jeong, H.Y.; Lee, C.; Kang, I.K.; Eom, S.H. Pigmentation and Flavonoid Metabolite Diversity in Immature 'Fuji' Apple Fruits in Response to Lights and Methyl Jasmonate. *Int. J. Mol. Sci.* **2022**, *23*, 1722. [CrossRef]
42. Jaakola, L. New insights into the regulation of anthocyanin biosynthesis in fruits. *Trends Plant Sci.* **2013**, *18*, 477–483. [CrossRef]
43. Xie, X.; Zhao, J.; Hao, Y.-J.; Fang, C.; Wang, Y. The ectopic expression of apple MYB1 and bHLH3 differentially activates anthocyanin biosynthesis in tobacco. *Plant Cell Tissue Organ Cult.* **2017**, *131*, 183–194. [CrossRef]

44. An, J.P.; Wang, X.F.; Li, Y.Y.; Song, L.Q.; Zhao, L.L.; You, C.X.; Hao, Y.-J. EIN3-LIKE1, MYB1, and ETHYLENE RESPONSE FACTOR3 Act in a Regulatory Loop That Synergistically Modulates Ethylene Biosynthesis and Anthocyanin Accumulation. *Plant Physiol.* **2018**, *178*, 808–823. [CrossRef]
45. An, J.P.; Liu, Y.J.; Zhang, X.W.; Bi, S.Q.; Wang, X.F.; You, C.X.; Hao, Y.J. Dynamic regulation of anthocyanin biosynthesis at different light intensities by the BT2-TCP46-MYB1 module in apple. *J. Exp. Bot.* **2020**, *71*, 3094–3109. [CrossRef]
46. Dandekari, A.M.; Teo, G.; Defilippi, B.G.; Uratsu, S.L.; Passey, A.J.; Kader, A.A.; Stow, J.R.; Colgan, R.J.; James, D.J. Effect of down-regulation of ethylene biosynthesis on fruit flavor complex in apple fruit. *Transgenic Res.* **2004**, *13*, 373–384. [CrossRef] [PubMed]
47. Schaffer, R.J.; Friel, E.N.; Souleyre, E.J.; Bolitho, K.; Thodey, K.; Ledger, S.; Bowen, J.H.; Ma, J.-H.; Nain, B.; Cohen, D.; et al. A genomics approach reveals that aroma production in apple is controlled by ethylene predominantly at the final step in each biosynthetic pathway. *Plant Physiol.* **2007**, *144*, 1899–1912. [CrossRef]
48. Wang, S.; Li, L.X.; Fang, Y.; Li, D.; Mao, Z.; Zhu, Z.; Chen, X.-S.; Feng, S.-Q. MdERF1B-MdMYC2 module integrates ethylene and jasmonic acid to regulate the biosynthesis of anthocyanin in apple. *Hortic. Res.* **2022**, *9*, uhac142. [CrossRef]
49. Wang, S.; Li, L.X.; Zhang, Z.; Fang, Y.; Li, D.; Chen, X.S.; Feng, S. Ethylene precisely regulates anthocyanin synthesis in apple via a module comprising MdEIL1, MdMYB1, and MdMYB17. *Hortic. Res.* **2022**, *9*, uhac034. [CrossRef]
50. Zhang, J.; Xu, H.; Wang, N.; Jiang, S.; Fang, H.; Zhang, Z.; Yang, G.; Wang, Y.; Su, M.; Xu, L.; et al. The ethylene response factor MdERF1B regulates anthocyanin and proanthocyanidin biosynthesis in apple. *Plant Mol. Biol.* **2018**, *98*, 205–218. [CrossRef] [PubMed]
51. Li, Y.Y.; Mao, K.; Zhao, C.; Zhao, X.Y.; Zhang, H.L.; Shu, H.R.; Hao, Y.J. MdCOP1 ubiquitin E3 ligases interact with MdMYB1 to regulate light-induced anthocyanin biosynthesis and red fruit coloration in apple. *Plant Physiol.* **2012**, *160*, 1011–1022. [CrossRef] [PubMed]
52. An, J.P.; Qu, F.J.; Yao, J.F.; Wang, X.N.; You, C.X.; Wang, X.F.; Hao, Y.-J. The bZIP transcription factor MdHY5 regulates anthocyanin accumulation and nitrate assimilation in apple. *Hortic. Res.* **2017**, *4*, 17023. [CrossRef] [PubMed]
53. Zhao, C.; Mao, K.; You, C.X.; Zhao, X.Y.; Wang, S.H.; Li, Y.Y.; Hao, Y.J. Molecular cloning and functional analysis of a UV-B photoreceptor gene, MdUVR8 (UV Resistance Locus 8), from apple. *Plant Sci.* **2016**, *247*, 115–126. [CrossRef] [PubMed]
54. Bi, X.; Zhang, J.; Chen, C.; Zhang, D.; Li, P.; Ma, F. Anthocyanin contributes more to hydrogen peroxide scavenging than other phenolics in apple peel. *Food Chem.* **2014**, *152*, 205–209. [CrossRef] [PubMed]
55. Chen, C.; Li, H.; Zhang, D.; Li, P.; Ma, F. The role of anthocyanin in photoprotection and its relationship with the xanthophyll cycle and the antioxidant system in apple peel depends on the light conditions. *Physiol. Plant.* **2013**, *149*, 354–366. [CrossRef] [PubMed]
56. Tossi, V.; Lamattina, L.; Cassia, R. An increase in the concentration of abscisic acid is critical for nitric oxide-mediated plant adaptive responses to UV-B irradiation. *New Phytol.* **2009**, *181*, 871–879. [CrossRef]
57. Zhang, J.; Chen, C.; Zhang, D.; Li, H.; Li, P.; Ma, F. Reactive oxygen species produced via plasma membrane NADPH oxidase regulate anthocyanin synthesis in apple peel. *Planta* **2014**, *240*, 1023–1035. [CrossRef]
58. Yin, H.; Wada, Y.; Kitamura, T.; Kambe, S.; Murasawa, S.; Mori, H.; Sakata, T.; Yanagida, S. Hydrothermal synthesis of nanosized anatase and rutile TiO_2 using amorphous phase TiO_2. *J. Mater. Chem.* **2001**, *11*, 1694–1703. [CrossRef]
59. Wang, Y.X.; Wang, X.J.; Cao, Y.; Zhong, M.S.; Zhang, J.; Yu, K.; Li, Z.-W.; You, C.-X.; Li, Y.-Y. Chemical composition and morphology of apple cuticular wax during fruit growth and development. *Fruit Res.* **2022**, *2*, 5. [CrossRef]
60. Wang, Y.; Zhang, J.; Wang, X.; Zhang, T.; Zhang, F.; Zhang, S.; Li, Y.; Gao, W.; You, C.; Wang, X.; et al. Cellulose Nanofibers Extracted From Natural Wood Improve the Postharvest Appearance Quality of Apples. *Front. Nutr.* **2022**, *9*, 881783. [CrossRef]
61. Lee, H.S.; Wicker, L. Anthocyanin Pigments in the Skin of Lychee Fruit. *J. Food Sci.* **2010**, *56*, 466–468. [CrossRef]
62. Li, Y.G.; Tanner, G.; Larkin, P. The DMACA–HCl Protocol and the Threshold Proanthocyanidin Content for Bloat Safety in Forage Legumes. *J. Sci. Food Agric.* **1996**, *70*, 89–101. [CrossRef]
63. Wang, Y.; Zhang, J.; Wang, D.; Wang, X.; Zhang, F.; Chang, D.; You, C.; Zhang, S.; Wang, X. Effects of cellulose nanofibrils treatment on antioxidant properties and aroma of fresh-cut apples. *Food Chem.* **2023**, *415*, 135797. [CrossRef] [PubMed]
64. Bolger, A.M.; Lohse, M.; Usadel, B. Trimmomatic: A flexible trimmer for Illumina sequence data. *Bioinformatics* **2014**, *30*, 2114–2120. [CrossRef] [PubMed]
65. Anders, S.; Huber, W. *Differential Expression of RNA-Seq Data at the Gene Level—The DESeq Package*; European Molecular Biology Laboratory (EMBL): Heidelberg, Germany, 2012.
66. Anders, S.; Pyl, P.T.; Huber, W. HTSeq—A Python framework to work with high-throughput sequencing data. *Bioinformatics* **2015**, *31*, 166–169. [CrossRef]
67. Kanehisa, M.; Araki, M.; Goto, S.; Hattori, M.; Hirakawa, M.; Itoh, M.; Katayama, T.; Kawashima, S.; Okuda, S.; Tokimatsu, T.; et al. KEGG for linking genomes to life and the environment. *Nucleic Acids Res.* **2007**, *36*, 480–484. [CrossRef]
68. Roberts, A.; Trapnell, C.; Donaghey, J.; Rinn, J.L.; Pachter, L. Improving RNA-Seq expression estimates by correcting for fragment bias. *Genome Biol.* **2011**, *12*, R22. [CrossRef] [PubMed]
69. Trapnell, C.; Williams, B.A.; Pertea, G.; Mortazavi, A.; Kwan, G.; van Baren, M.J.; Salzberg, S.L.; Wold, B.J.; Pachter, L. Transcript assembly and quantification by RNA-Seq reveals unannotated transcripts and isoform switching during cell differentiation. *Nat. Biotechnol.* **2010**, *28*, 511–515. [CrossRef]

70. Kim, D.; Langmead, B.; Salzberg, S.L. HISAT: A fast spliced aligner with low memory requirements. *Nat. Methods* **2015**, *12*, 357–360. [CrossRef]
71. Zhao, Y.; Zuo, J.; Yuan, S.; Shi, W.; Shi, J.; Feng, B.; Wang, Q. UV-C Treatment Maintains the Sensory Quality, Antioxidant Activity and Flavor of Pepino Fruit during Postharvest Storage. *Foods* **2021**, *10*, 2964. [CrossRef]
72. Dai, K.; Wei, Y.; Jiang, S.; Xu, F.; Wang, H.; Zhang, X.; Shao, X. Lab Scale Extracted Conditions of Polyphenols from Thinned Peach Fruit Have Antioxidant, Hypoglycemic, and Hypolipidemic Properties. *Foods* **2022**, *11*, 99. [CrossRef]
73. Kenneth, S. Spectrophotometric Method for Determination of Lipoxidase Activity 12. *Plant Physiol.* **1964**, *39*, 65. [CrossRef]

foods

MDPI

Article

A Dual-Signaling Electrochemical Aptasensor Based on an In-Plane Gold Nanoparticles–Black Phosphorus Heterostructure for the Sensitive Detection of Patulin

Jinqiong Xu [1], Jianbo Liu [1], Wuwu Li [1], Yongsheng Wei [1], Qinglin Sheng [2,3,*] and Yonghui Shang [1,*]

[1] College of Chemistry and Chemical Engineering, Xianyang Normal University, Xianyang 712000, China
[2] College of Food Science and Technology, Northwest University, Xi'an 710069, China
[3] Laboratory of Nutritional and Healthy Food-Individuation Manufacturing Engineering/Research Center of Food Safety Risk Assessment and Control, Xi'an 710069, China
* Correspondence: qlsheng@nwu.edu.cn (Q.S.); shangyh@xync.edu.cn (Y.S.)

Abstract: Patulin (PAT), a type of mycotoxin existing in foodstuffs, is harmful to food safety and human health. Thus, it is necessary to develop sensitive, selective and reliable analytical methods for PAT detection. In this study, a sensitive aptasensor based on a dual-signaling strategy was fabricated, in which a methylene-blue-labeled aptamer and ferrocene monocarboxylic acid in the electrolyte acted as a dual signal, for monitoring PAT. To improve the sensitivity of the aptasensor, an in-plane gold nanoparticles–black phosphorus heterostructure (AuNPs-BPNS) was synthesized for signal amplification. Due to the combination of AuNPs-BPNS nanocomposites and the dual-signaling strategy, the proposed aptasensor has a good analytical performance for PAT detection with the broad linear range of 0.1 nM–100.0 μM and the low detection limit of 0.043 nM. Moreover, the aptasensor was successfully employed for real sample detection, such as apple, pear and tomato. It is expected that BPNS-based nanomaterials hold great promise for developing novel aptasensors and may provide a sensing platform for food safety monitoring.

Keywords: black phosphorous; in-plane; electrochemical aptasensor; dual-signaling; patulin

1. Introduction

Patulin (a polyketide lactone 4-hydroxy-4H-furo(3,2c) pyran-2(6H)-one, PAT) belongs to a class of toxic compounds known as mycotoxins and is a common contaminant in food and feed. It is a toxic metabolite of at least 60 different species of fungi, such as *Penicillium expansum* (*P. leucopus*), *P. patulum* (*P. urticae, P. griseofulvum*), *P. crustosum* and *A. clavatus* [1]. Therefore, humans are likely to be exposed to such toxic substances, which may cause various toxic effects, such as lung and liver bleeding, kidney damage, neurotoxicity, teratogenicity and possible carcinogenicity [2]. Accordingly, the levels of PAT in food are controlled by a set of regulations. Marín et al. [2] reported that the maximum level of patulin in pears and apples is 126 mg/kg in the Spanish market, which is twice as high as the European Union (EU) standard. The World Health Organization and the Food and Agriculture Organization recommend a maximum daily level of PAT of 0.4 μg/kg body weight/day [3]. In China, patulin is also a common contaminant in a variety of fruits and products. The rate of patulin detection was 76.9% in 401 fruit products from eight provinces in China [4]. Hence, different analytical techniques have been explored for highly sensitive and accurate detection of PAT, such as chromatography [5–8], colorimetry [9] and immunoassays [10,11]. However, these methods required heavy instrumentation and complex technology, especially concerning the instability of antibodies in immunoassays [12]. Electrochemical sensors are a powerful technique because of their fast response time, low cost and excellent flexibility.

At present, aptamers, a type of DNA or RNA, can recognize various target molecules (metal ions, small molecules, proteins, etc.) with high affinity and have drawn interest in sensor design [13,14]. Combining aptamers and electrochemical methods, electrochemical aptasensors possess an excellent performance. However, most electrochemical aptasensors have only one signal readout: either a "signal off" or "signal on". The sensitivity of one-signal electrochemical sensors is inevitably limited, owing to signal fluctuation [15,16]. A dual-signaling strategy can solve these drawbacks. Numerous studies of electrochemical aptasensors with dual-signal channels have been reported to realize the high-performance detection of proteins and enzymes [17,18]. In dual-signaling electrochemical sensors, two electrochemical signals are dependent on targets, and their peak intensities show associated changes simultaneously. Therefore, dual-signaling electrochemical sensors exhibit a promising performance, with benefits such as a lower limit of detection and higher sensitivity.

To improve the sensitivity of electrochemical sensors, many nanomaterials have been applied for the amplification of signal. Among the various nanomaterials, two-dimensional (2D) nanomaterials have attracted widespread attention due to their unique physicochemical properties [19]. The first 2D nanomaterials appeared in 2004, when graphene was exfoliated from graphite by the mechanical cleavage technique [20,21]. Black phosphorus (BP), an allotrope of phosphorus with an orthorhombic lattice, is emerging as an important successor to 2D nanomaterials [22]. BP nanosheets (BPNS) can be exfoliated from bulk-layered crystal by breaking down its relatively weak van der Waals interactions, where each phosphorus atom forms covalent bonds with three neighboring atoms, forming a puckered orthorhombic structure [23,24]. The intriguing physical properties such as the high charge-carrier mobility, large on/off ratio, significant anisotropy and layer-dependent band gap have spurred tremendous interest in BP [25–28]. Owing to a direct, tunable band gap and no cytotoxicity, BPNS was used to explore photocatalyst and biomedical applications and more [29–31]. In order to improve the properties of BPNS and broaden their application, some nanocomposites based on BPNS have been synthesized by hybridization, doping or functionalization [32]. Sun et al. developed BPNS/graphene for sodium-ion batteries, which was self-assembled by intercalating BPNS between graphene layers via van der Waals interactions [33]. Hersam and co-workers reported that BPNS was functionalized by aryl diazonium through the formation of P–C covalent bonds and applied for field effect transistors [34]. Currently, the application of BPNS or BPNS-based nanocomposites in electrochemical sensors has not been adequately explored, which may be because of the instability of BP. Our group prepared gold nanoparticles–BPNS by electrostatic attraction for PAT electrochemical sensing [35]. However, this preparation method requires a more intensive experimental process and cannot enhance the stability of BP. The preparation of monodispersed BP nanosheets depends on long-time ultrasonic treatment or shear exfoliation, during which the BP layers are gradually cleaved and defects are generated on the edges and surface. These defects not only lead to the rapid degradation of the BP nanosheets, but also undermine the conductivity and electrochemical activity, thus causing problems in many applications—especially long-term catalysis. It is a good strategy to design the heterostructures to tailor the properties of 2D nanomaterials [36]. Due to the intrinsic advantages of the in-plane heterostructures, the in-plane charge carrier mobility of 2D materials is much higher than that between interlayers.

In this work, an in-plane gold nanoparticles–BPNS (AuNPs–BPNS) heterostructure was synthesized, in which the high chemical activity of the defect sites in BPNS can edge-selectively reduce $HAuCl_4$, and the formed AuNPs not only occupy the defect sites but also provide effective immobilization sites for the aptamer. Then, the obtained AuNPs–BPNS nanocomposites were used to construct a simple dual-signaling electrochemical aptasensor for sensitive and accurate detection of PAT (Scheme 1). The dual-signal was obtained from a methylene blue (MB)-labeled aptamer and ferrocene monocarboxylic acid (FMCA) in the electrolyte. The MB-labeled PAT aptamer self-assembled on the surface of pre-prepared AuNPs–BPNS/glassy carbon electrode (GCE) through the Au–S bond. Then, single-stranded DNA complementary to the aptamer formed a rigid double strand. In the presence of PAT, the aptamer recognized PAT and the complementary

DNA was released. As a result of the conformational change in the aptamer, the MB tag came close to the surface of the electrode and the diffusion of FMCA was hindered due to the increased steric resistance. Thus, the peak current of MB increased and FMCA decreased, producing "signal on" and "signal off". In this sensing system, because of the electronic property of BPNS and the sensitivity of the dual-signaling strategy, the proposed aptasensor exhibited an excellent performance for PAT detection. Moreover, the prepared aptasensor could also be applied in real sample detection. The developed sensing platform exhibited a great potential for the sensitive detection of other aptamer-specific binding contaminants of foodstuffs.

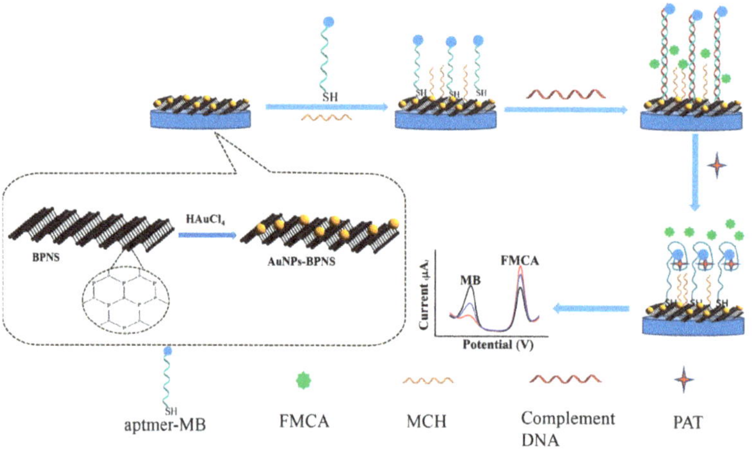

Scheme 1. Assembly diagram of the electrochemical dual-signaling aptasensor for PAT detection.

2. Materials and Methods

2.1. Chemicals and Reagents

The BP crystals were purchased from HWRK Chemical Co., Ltd. (Beijing, China) and stored in a glove box under dark. Chloroauric acid ($HAuCl_4$) was purchased from Acros Organics (Shanghai, China). *N*-methyl-2-pyrrolidone (NMP) was purchased from Aldrich (Shanghai, China). PAT, tris(2-carboxyethyl)phosphine hydrochloride solution (TCEP) and 6-mercaptohexanol (MCH) were purchased from Sangon Biotech. Co., Ltd. (Shanghai, China). All other solvents were of analytical grade. All solutions were prepared with ultrapure water (>18 MΩcm) from a Millipore Milli-Q water purification system.

The PAT aptamer and DNA sequence employed in this research were purchased from TaKaRa (Dalian, China). The sequences were as follows:

PAT aptamer:
5′-MB-CCCGGCCCGCCAACCCGCATCATCTACACTGATATTTTACCTTCCC-SH-3′.
Complement DNA (cDNA): 5′-AAAGGGTTGGCGGGCC-3′.

2.2. Apparatus and Characterization

All of the electrochemical experiments were performed with an IGS1130 electrochemical workstation (Guangzhou Insens Sensor Technology Co. Ltd, Guangzhou, China). The three-electrode system consisted of a modified GCE (3 mm in diameter) as the working electrode, a platinum wire as the auxiliary electrode, and Ag/AgCl (saturated KCl) electrode as the reference electrode; this system was used in all electrochemical investigations. Transmission electron microscopy (TEM) images were acquired with a Hitachi H-7500 high-resolution transmission electron microscope (Tokyo, Japan) using an accelerating voltage of 200 kV. UV–Vis absorption spectra were acquired on a Cary 50 Scan UV–Vis spectrophotometer (Varian, Australia). X-ray photoelectron spectra (XPS) were obtained with an ESCALAB 250 Xi XPS system from Thermo Scientific (Waltham, MA, USA). Raman spectra were recorded using a Renishaw Micro-Raman spectroscopy system (London, England) with an excitation wavelength of 514 nm.

2.3. Preparation of BPNS

For preparation of BPNS, 5 mg of BP crystal was dispersed in 50 mL saturated sodium hydroxide solution of NMP under sonication at room temperature for 12 h. Then, the solution was centrifuged at 4000 rpm for 10 min and the supernatant was collected. Finally, BPNS were obtained by centrifuging the supernatant at 12,000 rpm for 15 min and dispersed in water as a brown solution.

2.4. Preparation of AuNPs–BPNS

AuNPs were modified on BPNS by in situ reduction. Briefly, 250 μL of 25 mM $HAuCl_4$ was added to 1 mL of 1 mg/mL BPNS solution and stirred for 1 h in oxygen-free dark conditions. The products were separated by centrifugation and washed with deionized water. Finally, the obtained product was dispersed in water and stored at 4 °C.

2.5. Preparation of the Modified Electrodes

The GCE was polished to a mirror-like surface by 1.0 and 0.3 μm alumina slurries, respectively, and completely cleaned in ethanol and ultrapure water by ultrasonication. Then, 10 μL of a 1 mg/mL AuNPs–BPNS solution was dropped into GCE and dried in air at room temperature. Before the aptamer was fixed, the aptamer was added to phosphate-buffered saline solution (PBS; pH 7.0) containing 10 mM TCEP for treatment for 1 h to reduce the presence of disulfide bonds, and then the treated aptamer was dropped onto AuNPs–BPNS/GCE for 12 h. The electrode was then immersed in 1 mM MCH for 1 h to block the unbound sites. Finally, 10 μL of 1 μM cDNA was dropped onto the electrode surface for 1.5 h to obtain the final modified electrode cDNA/aptamer/AuNPs–BPNS/GCE.

2.6. Electrochemical Detection of PAT

For the detection of PAT, the modified electrode was immersed in 20 μL of a fixed concentration of PAT for 60 min under room temperature and then rinsed with the PBS to wash away the unbound PAT. The obtained electrode was then immersed in PBS (pH 7.4) containing 1 μM FMCA and investigated by cyclic voltammetry (CV) and cyclic voltammetry (SWV). The selective experiments for hydroxymethylfurfural (HMF), aflatoxin B1 (AFB1), aflatoxin M1 (AFM), ochratoxin A (OTA) and zearalenone (ZEN) were performed under the same conditions. All experiments were performed at room temperature and measured three times.

2.7. Pretreatment of Food Samples

In this experiment, fresh and infected apples, pears, and tomatoes acted as real food samples. The pretreatment procedures were performed according to previous work [37]. Apples, pears and tomatoes were smashed separately, then 2.5 g of the obtained sample mash was added into 10 mL H_2O and extracted by ultrasonication for 1 h. The supernatant from each sample was transferred to a centrifuge tube and centrifuged at 6000 rpm for 5 min. Then, the extracts were collected. Finally, the supernatant and different concentrations of PAT were added to PBS.

3. Results

3.1. Characterization of the AuNPs–BPNS

The morphologies of the as-prepared BPNS and AuNPs–BPNS were characterized by TEM technologies. The TEM micrograph of BPNS is presented in Figure 1A. BPNS presents a smooth surface with some wrinkles, which is a 2D nanostructure. Figure 1B shows the TEM patterns of AuNPs–BPNS nanocomposites. AuNPs were successfully synthesized on the surface of BPNS without other reduction and the inset of Figure 1B shows an average diameter of AuNPs of about 21 nm. Elemental analyses of the AuNPs–BPNS nanocomposites entailed EDX mappings of P, Au and O, where P and O originated from BPNS and Au indicated the presence of AuNPs (Figure 1C). Notably, AuNPs were uniformly distributed on the entire BPNS surface. As shown in Figure 1D, the EDX analysis further revealed the existence of P, Au and O in the nanocomposites.

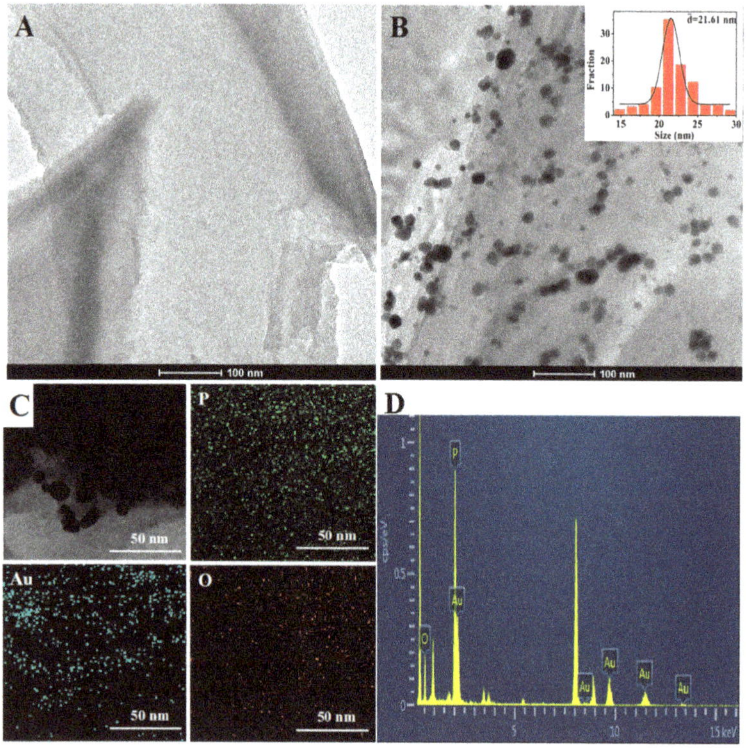

Figure 1. TEM images of BPNS (**A**) and AuNPs–BPNS (**B**). Inset: the size distributions of AuNPs. EDX mappings of P, Au and O in the AuNPs–BPNS (**C**). EDX spectrum of AuNPs–BPNS (**D**).

The effect of $HAuCl_4$ concentration on the morphology of AuNPs–BPNS was also investigated. As shown in Figure 2A, when 10 mM $HAuCl_4$ was added, only a small amount of AuNPs were generated on the surface of the BPNS. As the $HAuCl_4$ concentration increased to 20 mM (Figure 2B), the amount of AuNPs on the BPNS surface increased. A satisfying amount of AuNPs–BPNS could be obtained once the concentration of $HAuCl_4$ was increased to 25 mM (Figure 2C). It can be seen that AuNPs were uniformly deposited on the surface of BPNS without obvious aggregation, which may be because the defect of BPNS is exactly occupied by the AuNPs. However, when 30 mM $HAuCl_4$ was added (Figure 2D), the AuNPs showed obvious aggregation. Therefore, 25 mM $HAuCl_4$ was selected to prepare the AuNPs–BPNS nanocomposites.

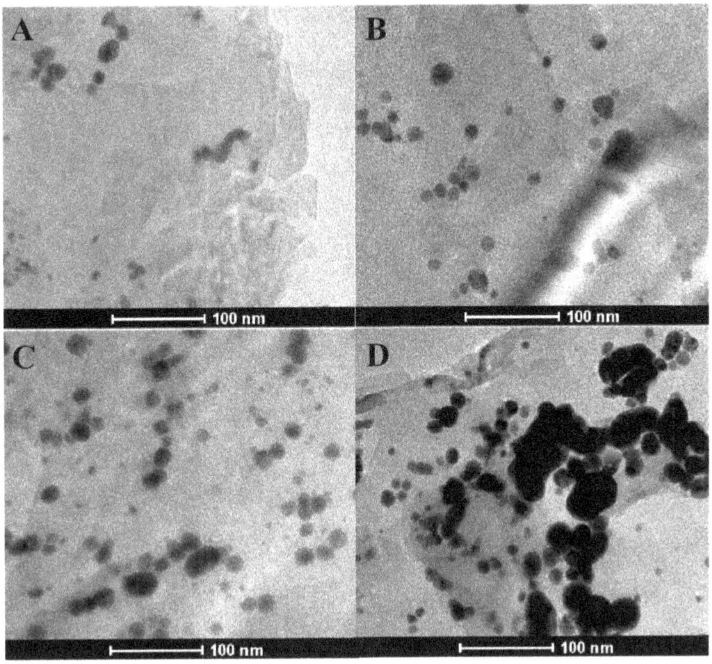

Figure 2. TEM images of AuNPs–BPNS synthesized while adding different concentrations of $HAuCl_4$: (**A**) 10 mM; (**B**) 20 mM; (**C**) 25 mM; (**D**) 30 mM.

UV–Vis spectroscopy was conducted to characterize the formation process of the AuNPs. Figure 3A shows the UV–Vis spectrum of BPNS with different amounts of added $HAuCl_4$. As the amount of BPNS increased, the absorption peaks of the AuNPs increased at 536 nm, demonstrating that the formation of the AuNPs is related to the BPNS. Additionally, the UV–Vis spectrum of $HAuCl_4$ added to BPNS with the increase in mixing time is shown in Figure 3B. The absorption peak intensity of the AuNPs increased with the increase in time, and the color of the solution changed to different shades of a red color. The successful synthesis of AuNPs may be due to the edge-selective reduction of BPNS. Moreover, AuNPs can occupy the defects in BPNS to enhance the stability of BPNS [38]. Raman spectroscopy could provide detailed vibrational and rotational patterns for sample identification. Raman spectra of BPNS and AuNPs–BPNS are provided in Figure 3C. Three peaks are displayed at 363 cm^{-1}, 440 cm^{-1} and 466 cm^{-1}, which were ascribed to one out-of-plane phonon mode $A_g{}^1$ and two in-plane phonon modes B_{2g} and $A_g{}^2$ for BP, respectively [39,40]. Upon AuNPs' modification, the three peaks blue-shifted slightly because of vibration inhibition. Particularly, the B_{2g} and $A_g{}^2$ mode shift was larger than that of the $A_g{}^1$ mode, indicating that AuNPs–BPNS possess an in-plane heterostructure. Furthermore, the changes in oxidation states of BPNS from pristine BPNS to AuNPs–BPNS were probed by XPS (Figure S1). In Figure 3D, curves a and b display the P2p spectra before and after AuNP modification, respectively. The two unchanged peaks near 128.5 and 129.5 eV verify the original state of BP, which is consistent with standard BP. The peak at 133 eV is due to oxidized phosphorus (i.e., PO_x) [41–43]. However, for AuNPs–BPNS (Figure 3Db), the PO_x peak was enhanced, demonstrating that some BP was oxidized in the formation of AuNPs–BPNS. The experimental data coincide with those reported in the literature [38].

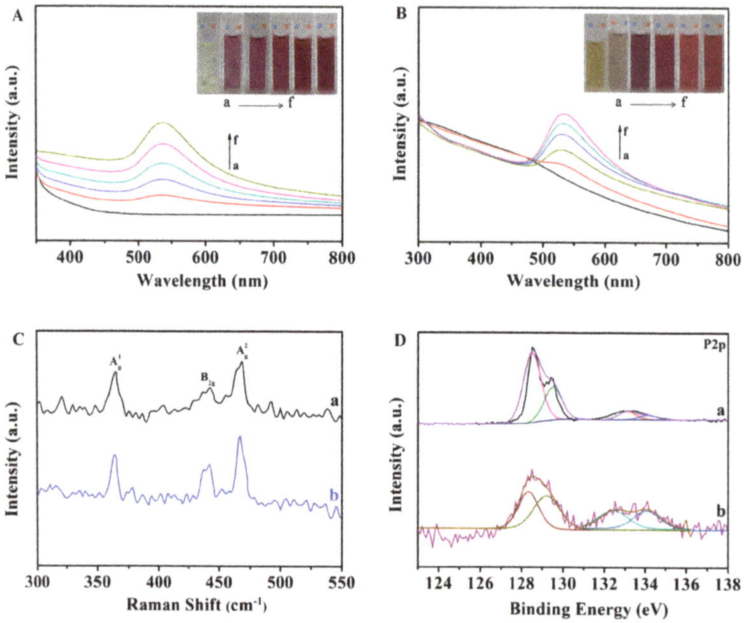

Figure 3. (**A**) UV–Vis spectrum of $HAuCl_4$ mixed with BPNS at different amounts (from a to f: 0, 0.2, 0.4, 0.6, 0.8, 1 mg/mL). (**B**) UV–Vis spectrum of $HAuCl_4$ added to BPNS for different times (from a to f: 0, 15, 30, 40, 50, 60 min). (**C**) Raman spectra of BPNS (curve a) and AuNPs–BPNS (curve b). (**D**) High-resolution XPS spectra of the P2p core level for BPNS (curve a) and AuNPs–BPNS (curve b).

3.2. Characterization of the Modified Electrode

EIS is an efficient method for exploring the interfacial properties of the modified electrodes [44,45]. Therefore, the modification process was characterized by EIS to reveal the electrodes' surface features. EIS measurement was performed in 5.0 mM $K_3[Fe(CN)_6]/K_4[Fe(CN)_6]$ (1:1) solution containing 0.1 M KCl. The equipment was set at 5 mV of alternative voltage amplitude and 200 mV of applied potential, and the voltage frequencies ranged from 0.01 Hz to 100 kHz. The electron transfer resistance (R_{et}) was expressed by the diameter of a semicircle. As shown in Figure 4A, the bare GCE showed a small semicircle diameter (177 Ω, curve a), indicating a fast electron-transfer process. When BPNS was immobilized on the GCE, the semicircle diameter became bigger than that of bare GCE (704 Ω, curve b). As a result of the good electron transfer efficiency of the AuNPs, the AuNPs–BPNS/GCE exhibited a lower R_{et} (566 Ω, curve c) than BPNS/GCE, suggesting that AuNPs could efficiently enhance the electron transfer efficiency. After the assembly of the aptamer, the value of R_{et} increased to 1123 Ω (curve d) because the negatively charged aptamer and $[Fe(CN)_6]^{3-/4-}$ anions repelled each other. With the hybridization between the aptamer and cDNA, the semicircle diameter further increased (curve e), which suggests the successful recognition of the aptamer and cDNA. The results of EIS verified that the aptasensor was successfully prepared (Scheme 1). Furthermore, XPS spectra for the C, N, Au and S energy regions are shown in Figure S2, which also verified the aptamer immobilization on AuNPs–BPNS.

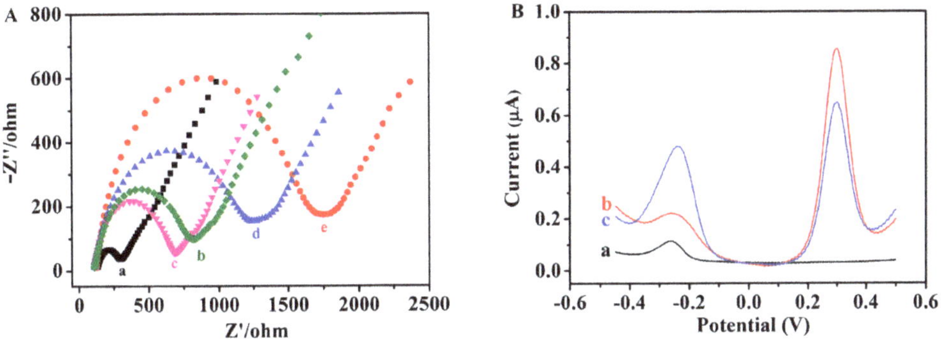

Figure 4. (**A**) EIS spectra of (a) the bare GCE, (b) BPNS/GCE, (c) AuNPs–BPNS/GCE, (d) aptamer/AuNPs–BPNS/GCE and (e) cDNA/aptamer/AuNPs–BPNS/GCE in a mixture solution of 5.0 mM $Fe(CN)_6{}^{3-/4-}$ containing 0.1 M KCl. (**B**) SWV measurements of cDNA/aptamer/AuNPs–BPNS/GCE in PBS (a) without FMCA, (b) with FMCA and (c) after incubation with PAT.

The SWV measurements were carried out to investigate the electrochemical behavior of cDNA/aptamer/AuNPs–BPNS/GCE in different conditions. As shown in Figure 4B, only a low oxidation peak of MB at −0.25 V (curve a) can be observed in PBS without FMCA. After the addition of FMCA into PBS, a significant oxidation peak was produced at 0.3 V (curve b). However, after incubation with PAT, the MB response increased significantly and the FMCA response decreased (curve c). The results reveal that the proposed aptasensor could be used for PAT detection. PAGE analysis was performed to demonstrate the DNA combination processes step by step. As shown in Figure S3 (Supporting Information), Lanes 1 and 2 correspond to the aptamer and cDNA, respectively. Lane 3 represents the mixture of aptamer and cDNA; a new band was observed, demonstrating the hybridization among aptamer and cDNA. After PAT was added to this mixture, the band of hybridization weakened but the band of cDNA was obvious, which can be attributed to the initiation of hybridization and the release of cDNA. It implies that the aptamer and cDNA hybridized successfully and the aptamer could recognize PAT; thus, the cDNA was released.

3.3. Electrochemical PAT Assay

In order to obtain an excellent performance in sensing, some parameters were optimized: (a) the concentration of aptamer, (b) the hybridization time of aptamer and cDNA, (c) the incubation time of PAT. Respective data and figures are given in the electronic Supporting Material (Figure S3). The following experimental conditions were found to give the best results: (a) optimal aptamer concentration: 1 μM; (b) optimal hybridization time: 90 min; (c) optimal incubation time: 60 min.

The SWV was implemented to study the analytical performance of the proposed aptasensor. As shown in Figure 5A, a series of different concentrations of PAT from 0.1 nM to 100 μM were measured in PBS containing 10 μM FMCA. The peak current of MB increased and that of FMCA decreased with increasing PAT concentration. Figure 5B displays the linear relationship between the individual MB or FMCA current response and the logarithmic values of PAT concentration. For the MB signal, the limit of detection (LOD) was 0.011 ng/mL with R = 0.9987, and for the FMCA signal, 0.012 ng/mL with R = 0.9967. Furthermore, from Figure 5C, the linear relationship between the value of I_{MB}/I_{FMCA} and the logarithm of the PAT concentration was acquired as follows: $I_{MB}/I_{FMCA} = 0.12 \log C_{PAT} + 0.38$ with R = 0.9941. A good linear correlation was obtained from 0.0154 ng/mL to 15.4 μg/mL and the LOD reached 0.066 ng/mL (S/N = 3). Compared with either an MB or FMCA signal sensor, the dual-signal strategy has a lower LOD.

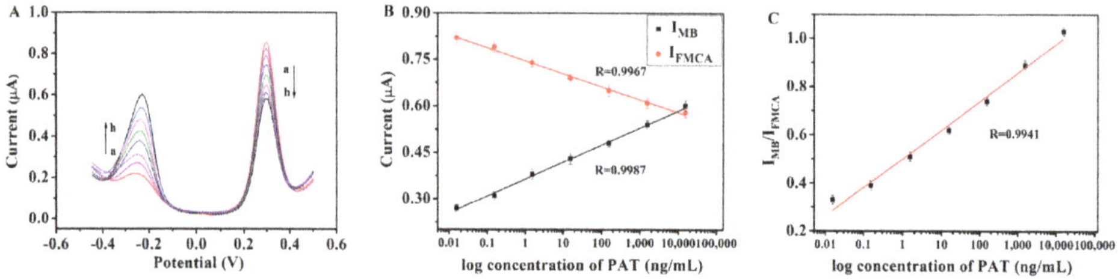

Figure 5. (**A**) SWV curve of the aptasensor at different PAT concentrations (from curve a to h: 0, 0.0154, 0.154, 1.54, 15.4, 154, 1540 and 15,400 ng/mL) in PBS containing 10 μM FMCA. (**B**) Calibration curve of current of MB and FMCA for the detection of PAT. (**C**) Linear relationship between I_{MB}/I_{FMCA} and logarithm of the PAT concentration. The error bars were calculated on the basis of the standard deviation of three independent experiments.

Moreover, the analytical performance for the proposed dual-signaling aptasensor is comparable with other methods in previous reports as can be seen in Table 1. This may be because BPNS has an excellent electronic performance and two-dimensional sheet structure with a large surface area which provides more sites for the generation of AuNPs and prevents AuNPs aggregation; in addition, the dual-signal ratio strategy greatly reduces the impact of the external environment on the detection results and improves the sensitivity of the sensor. The results indicate that the proposed aptasensor exhibits a lower LOD and wider linear range.

Table 1. Comparison of different methods for the determination of PAT.

Methos	Linear Range (ng/mL)	LOD (ng/mL)	Reference
Colorimetric detection	0.046–24.64	0.046	[46]
Florescent detection	15.4–1540	9.24	[47]
Gas chromatography-Mass spectrometry	64.9–1623	5.00	[48]
Fluorescent resonant energy transfer	0.0092–92.4	0.003	[49]
Quartz crystal microbalance immunoassay	30.8–154	21.56	[50]
High performance chromatography	0.92–400.4	0.154	[51]
Phosphorescent nanosensor	66.2–1001	49.28	[52]
Electrochemical detection	0.0154–15.4	0.043	This work

The linearity in real spiked matrix samples was determined using pretreated fresh apple juice containing 10 μM FMCA and different PAT concentrations. The calibration curve (Figure 6) was obtained from the linear regression of a plot of peak area versus concentration. Calibration was carried out in the range of 0.154–15,400 ng/mL, being found to be linear in the entire studied range with R = 0.9956.

3.4. Reproducibility, Stability and Selectivity

The reproducibility of the proposed aptasensor was evaluated by using five cDNA/aptamer/AuNPs–BPNS/GCEs for the detection of 0.154 μg/mL PAT under the same conditions (Figure 7A). The relative standard deviation (RSD) was calculated to be 4.5%, revealing significant reproducibility. Moreover, the aptasensor was stored under refrigeration for ten days. As shown in Figure 7B, 91.1% of the initial response of the current was obtained after adding the same amount of PAT (0.154 μg/mL), demonstrating the satisfactory stability of the proposed aptasensor.

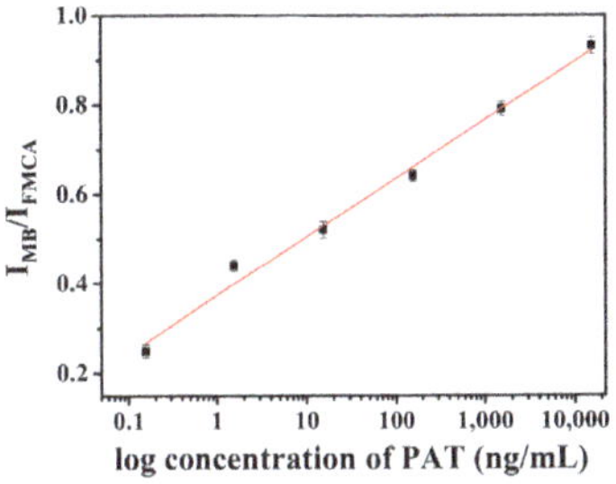

Figure 6. Linear relationship between I_{MB}/I_{FMCA} and the logarithm of PAT concentration in real spiked matrix samples. The error bars were calculated on the basis of the standard deviation of three independent experiments.

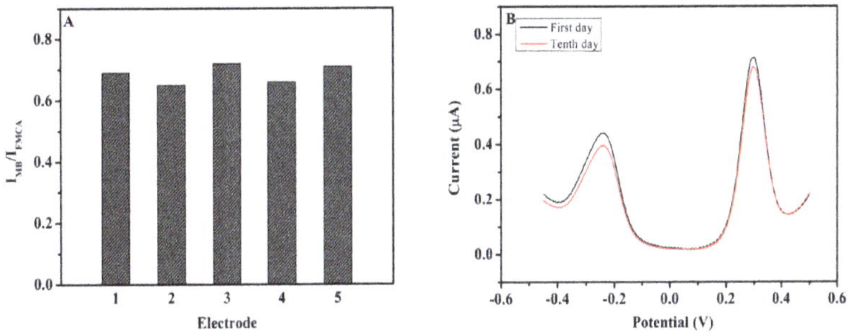

Figure 7. (**A**) Reproducibility of the aptasensor for 0.154 μg/mL PAT detection. The numbers 1–5 represent five independent electrodes, respectively. (**B**) Stability of the aptasensor for 0.154 μg/mL PAT detection.

Furthermore, five kinds of contaminants in food samples containing hydroxymethylfurfural (HMF), aflatoxin B1 (AFB1), aflatoxin M1 (AFM), ochratoxin A (OTA) and zearalenone (ZEN) were used to verify the selectivity of the aptasensor. As shown in Figure 8, despite the fact that the concentrations of the interferences were as high as 10 μM, there were low values of I_{MB}/I_{FMCA}. However, an obvious response was obtained by the addition of PAT. The results indicated that the proposed dual-signaling aptasensor exhibits an excellent selectivity for PAT.

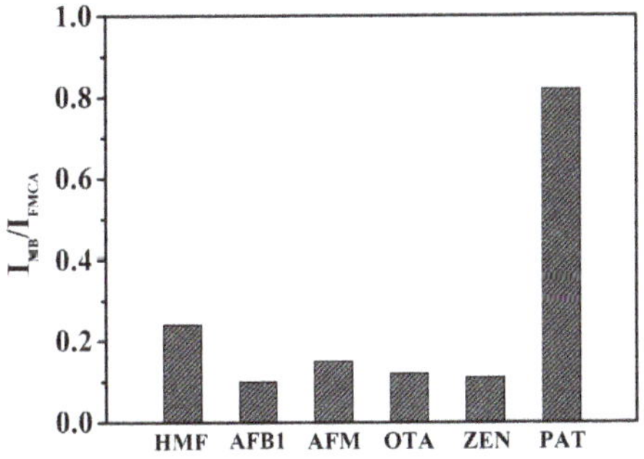

Figure 8. Selectivity of the PAT aptasensor. The concentration of AFB1, AFM, OTA, ZEN and PAT are 10 μM. The error bars were calculated on the basis of the standard deviation of three independent experiments.

3.5. Real Sample Detection

To evaluate the practical application capability of the present system, the proposed aptasensor was employed to detect the PAT contents in fresh and infected apple, pear, and tomato. As shown in Table 2, no PAT was found in the fresh foods, while small amounts were found in the infected foods. Then, the standard addition method was also performed. The standard PAT solutions were added into the food samples; the recoveries and RSDs were 95.0–106.7% and 2.9–5.1%. Furthermore, there was no significant difference between the results obtained by the aptasensor and an HPLC method, as shown in Table 1. These results implied that the dual-signaling aptasensor is applicable for PAT detection in food samples.

Table 2. Results of the determination of PAT in food samples and recovery tests.

Samples	Background Concentration (μg/mL)	Added Concentration (μg/mL)	Found Concentration (μg/mL, HPLC)	Found Concentration (μg/mL, the Sensor)	Recovery (%)	RSD (%, n = 3)
Fresh apple	ND	0.092	0.075	0.089	96.7	3.31
Fresh pear	ND	0.092	0.077	0.088	95.0	2.92
Fresh tomato	ND	0.092	0.088	0.094	101.7	4.05
Infected apple	0.035	0.092	0.104	0.131	103.3	4.31
		0.308	0.298	0.335	97.5	3.12
Infected pear	0.032	0.092	0.099	0.131	106.7	4.56
		0.308	0.301	0.344	101.5	3.60
Infected tomato	0.071	0.092	0.113	0.159	96.7	5.14
		0.308	0.315	0.374	98.5	3.52

ND: Not Detected.

4. Conclusions

In summary, BPNS have potential value in the field of food safety detection. Metal nanoparticles, especially AuNPs, as an efficient catalyst, promote electron transfer on the AuNPs–BPNS, overcoming the defects of single metal nanoparticles. In this paper, in-plane gold nanoparticles–BPNS (AuNPs–BPNS) was synthesized, in which the high chemical activity of the defect sites in BPNS can edge-selective reduced $HAuCl_4$, and the formed AuNPs not only enhance the stability of BPNS but also provide effective immobilization sites of the aptamer. The obtained AuNPs–BPNS nanocomposites were used to construct a dual-signaling aptasensor for PAT detection. Under optimum conditions, the proposed aptasensor was linear over the range of 0.0154 ng/mL–15.4 μg/mL. The electrochemical sensing platform has been successfully applied to the determination of PAT in apple, pear, and tomato extracts with good accuracy and reliability. Food safety detection methods based on functional nanomaterials will usher in new development opportunities in the future.

Supplementary Materials: The following supporting information can be downloaded at: https://www.mdpi.com/article/10.3390/foods12040846/s1, Figure S1: High-resolution XPS spectra for (A) C 1s, (B) N 1s, (C) Au 4f, and (D) S 2p regions collected from aptamer compound AuNPs-BPNS.; Figure S2: PAGE images of different DNAs. Lane 1, aptamer; Lane 2, cDNA; Lane 3, hybridization of aptamer and cDNA; Lane 4, hybridization of aptamer and cDNA treated with PAT.; Figure S3: Optimization of the concentration of aptamer (A), the hybridization time of aptamer and cDNA (B) and the incubation time of PAT (C) for the PAT (154 ng/mL) detection.

Author Contributions: J.X.: conceptualization, methodology, formal analysis, writing—original draft; J.L.: revision and formal analysis; W.L.: revision and formal analysis; Y.W.: revision and formal analysis; Q.S.: supervision, funding acquisition, review and editing; Y.S.: supervision. All authors have read and agreed to the published version of the manuscript.

Funding: The authors gratefully acknowledge the financial support of this project by the National Key Research and Development Program of China (No. 2019YFC1606703), the Natural Science Foundation of Shaanxi Province in China (No. 2020JM-429), the scientific research program funded by Shaanxi Provincial Education Department (No. 22JK0599), the Xianyang City Qinchuangyuan Science and Technology Innovation Special Project (L2022-QCYZX-GY-006) and the Natural Science Foundation of Xianyang Normal University (XSYK20003).

Data Availability Statement: The data are available from the corresponding authors.

Conflicts of Interest: The authors declare no conflict of interest.

References

1. Bennett, J.; Klich, M. Clinical Microbiology Reviews. *Mycotoxins* **2003**, *16*, 497.
2. Marín, S.; Mateo, E.M.; Sanchis, V.; Valle-Algarra, F.M.; Ramos, A.J.; Jiménez, M. Patulin contamination in fruit derivatives, including baby food, from the Spanish market. *Food Chem.* **2011**, *124*, 563–568. [CrossRef]
3. Leggott, N.L.; Shephard, G.S. Patulin in South African commercial apple products. *Food Control.* **2001**, *12*, 73–76. [CrossRef]
4. Wu, N.; Wang, Y.; Liu, Y.; Song, J.; Yang, X.; Su, C.; Liu, X. Investigation on the content of penicillin in fruit products in some areas of China. *Health Res.* **1992**, *2*, 101–103.
5. Veprikova, Z.; Vaclavikova, M.; Lacina, O.; Dzuman, Z.; Zachariasova, M.; Hajslova, J. Occurrence of mono- and di-glycosylated conjugates of T-2 and HT-2 toxins in naturally contaminated cereals. *World Mycotoxin J.* **2012**, *5*, 231–240. [CrossRef]
6. Seo, M.; Kim, B.; Baek, S.-Y. An optimized method for the accurate determination of patulin in apple products by isotope dilution-liquid chromatography/mass spectrometry. *Anal. Bioanal. Chem.* **2015**, *407*, 5433–5442. [CrossRef]
7. Yang, J.Z.; Wei, W.; Zhu, W.X.; Liu, Y.F. Extraction and preliminary purification of cochineal wax. *Food Sci.* **2009**, *30*, 162–165.
8. Kharandi, N.; Babri, M.; Azad, J. A novel method for determination of patulin in apple juices by gc-ms. *Food Chem.* **2013**, *141*, 1619–1623. [CrossRef]
9. Lu, C.; Chen, X.; Liu, C.; Lu, C. Development and validation of a label-free colorimetric aptasensor based on the HCR and hemin/G-quadruplex DNAzyme for the determination of patulin in fruits and fruit-based products from Xinjiang (China). *Anal. Methods* **2022**, *14*, 3375–3381. [CrossRef]
10. Pennacchio, A.; Varriale, A.; Esposito, M.G.; Staiano, M.; D'Auria, S. A near-infrared fluorescence assay method to detect patulin in food. *Anal. Biochem.* **2015**, *481*, 55–59. [CrossRef]
11. Pennacchio, A.; Ruggiero, G.; Staiano, M.; Piccialli, G.; Oliviero, G.; Lewkowicz, A.; Synak, A.; Bojarski, P.; D'Auria, S. A surface plasmon resonance based biochip for the detection of patulin toxin. *Opt. Mater.* **2014**, *36*, 1670–1675. [CrossRef]
12. Zhao, Y.; Yang, Y.; Luo, Y.; Yang, X.; Li, M.; Song, Q. Double Detection of Mycotoxins Based on SERS Labels Embedded Ag@Au Core–Shell Nanoparticles. *ACS Appl. Mater. Interfaces* **2015**, *7*, 21780–21786. [CrossRef]
13. Esteban-Fernández de Ávila, B.; Lopez-Ramirez, M.A.; Ba, D.F.; Jodra, A.; Singh, V.V.; Kaufmann, K.; Wang, J. Aptamer-modified graphene-based catalytic micromotors: Off–on fluorescent detection of ricin. *ACS Sens.* **2016**, *1*, 217–221. [CrossRef]
14. Cui, M.; Wang, Y.; Jiao, M.; Jayachandran, S.; Wu, Y.; Fan, X.; Luo, X. Mixed Self-Assembled Aptamer and Newly Designed Zwitterionic Peptide as Antifouling Biosensing Interface for Electrochemical Detection of alpha-Fetoprotein. *ACS Sens.* **2017**, *2*, 490–494. [CrossRef]
15. Du, Y.; Lim, B.J.; Li, B.; Jiang, Y.S.; Sessler, J.L.; Ellington, A.D. Reagentless, Ratiometric Electrochemical DNA Sensors with Improved Robustness and Reproducibility. *Anal. Chem.* **2014**, *86*, 8010–8016. [CrossRef]
16. Xiong, E.; Li, Z.; Zhang, X.; Zhou, J.; Yan, X.; Liu, Y.; Chen, J. Triple-Helix Molecular Switch Electrochemical Ratiometric Biosensor for Ultrasensitive Detection of Nucleic Acids. *Anal. Chem.* **2017**, *89*, 8830–8835. [CrossRef]
17. Meng, F.; Chen, X.; Cheng, W.; Hu, W.; Tang, Y.; Miao, P. Ratiometric Electrochemical Sensing Strategy for the Ultrasensitive Detection of Telomerase Activity. *Chemelectrochem* **2019**, *6*, 2000–2003. [CrossRef]
18. Ren, K.; Wu, J.; Yan, F.; Zhang, Y.; Ju, H. Immunoreaction-triggered DNA assembly for one-step sensitive ratiometric electrochemical biosensing of protein biomarker. *Biosens. Bioelectron.* **2015**, *66*, 345–349. [CrossRef]
19. Geim, A.K.; Novoselov, K.S. The rise of graphene. *Nat. Mater.* **2007**, *6*, 183–191. [CrossRef]
20. Novoselov, K.S.; Geim, A.K.; Morozov, S.V.; Jiang, D.; Zhang, Y.; Dubonos, S.V.; Grigorieva, I.V.; Firsov, A.A. Electric field effect in atomically thin carbon films. *Science* **2004**, *306*, 666–669. [CrossRef]
21. Zhang, H. Ultrathin Two-Dimensional Nanomaterials. *ACS Nano* **2015**, *9*, 9151–9469. [CrossRef] [PubMed]
22. Eswaraiah, V.; Zeng, Q.; Long, Y.; Liu, Z. Black phosphorus nanosheets: Synthesis, characterization and applications. *Small* **2016**, *12*, 3480–3502. [CrossRef] [PubMed]
23. Mao, N.; Tang, J.; Xie, L.; Wu, J.; Han, B.; Lin, J.; Deng, S.; Ji, W.; Xu, H.; Liu, K.; et al. Optical Anisotropy of Black Phosphorus in the Visible Regime. *J. Am. Chem. Soc.* **2015**, *138*, 300–305. [CrossRef] [PubMed]
24. Mao, N.; Wu, J.; Han, B.; Lin, J.; Tong, L.; Zhang, J. Birefringence-Directed Raman Selection Rules in 2D Black Phosphorus Crystals. *Small* **2016**, *12*, 2627–2633. [CrossRef]
25. Rahman, M.Z.; Kwong, C.W.; Davey, K.; Qiao, S.Z. 2D phosphorene as a water splitting photocatalyst: Fundamentals to applications. *Energy Environ. Sci.* **2016**, *9*, 709–728. [CrossRef]

26. Liu, H.; Du, Y.; Deng, Y.; Peide, D.Y. Semiconducting black phosphorus: Synthesis, transport properties and electronic applications. *Chem. Soc. Rev.* **2015**, *46*, 2732–2743. [CrossRef]
27. Li, L.; Kim, J.; Jin, C.; Ye, G.J.; Qiu, D.Y.; da Jornada, F.H.; Shi, Z.; Chen, L.; Zhang, Z.; Yang, F.; et al. Direct observation of the layer-dependent electronic structure in phosphorene. *Nat. Nanotechnol.* **2016**, *12*, 21–25. [CrossRef]
28. Lei, W.; Zhang, T.; Gu, L.; Liu, P.; Rodriguez, J.A.; Liu, G.; Liu, M. Surface-Structure Sensitivity of CeO_2 Nanocrystals in Photocatalysis and Enhancing the Reactivity with Nanogold. *ACS Catal.* **2015**, *5*, 4385–4393. [CrossRef]
29. Hu, J.; Guo, Z.; Mcwilliams, P.E.; Darges, J.E.; Druffel, D.L.; Moran, A.M.; Warren, S.C. Band Gap Engineering in a 2D Material for Solar-to-Chemical Energy Conversion. *Nano Lett.* **2015**, *16*, 74–79. [CrossRef]
30. Wang, H.; Yang, X.; Shao, W.; Chen, S.; Xie, J.; Zhang, X.; Wang, J.; Xie, Y. Ultrathin Black Phosphorus Nanosheets for Efficient Singlet Oxygen Generation. *J. Am. Chem. Soc.* **2015**, *137*, 11376–11382. [CrossRef]
31. Ren, X.; Li, Z.; Huang, Z.; Sang, D.; Qiao, H.; Qi, X.; Li, J.; Zhong, J.; Zhang, H. Environmentally Robust Black Phosphorus Nanosheets in Solution: Application for Self-Powered Photodetector. *Adv. Funct. Mater.* **2017**, *27*, 1606834. [CrossRef]
32. Lei, W.; Liu, G.; Zhang, J.; Liu, M. Black phosphorus nanostructures: Recent advances in hybridization, doping and functionalization. *Chem. Soc. Rev.* **2017**, *46*, 3492–3509. [CrossRef]
33. Sun, J.; Lee, H.-W.; Pasta, M.; Yuan, H.; Zheng, G.; Sun, Y.; Li, Y.; Cui, Y. A phosphorene–graphene hybrid material as a high-capacity anode for sodium-ion batteries. *Nat. Nanotechnol.* **2015**, *10*, 980–985. [CrossRef]
34. Ryder, C.R.; Wood, J.D.; Wells, S.A.; Yang, Y.; Jariwala, D.; Marks, T.J.; Schatz, G.C.; Hersam, M.C. Covalent functionalization and passivation of exfoliated black phosphorus via aryl diazonium chemistry. *Nat. Chem.* **2016**, *8*, 597–602. [CrossRef]
35. Xu, J.; Qiao, X.; Wang, Y.; Sheng, Q.; Yue, T.; Zheng, J.; Zhou, M. Electrostatic assembly of gold nanoparticles on black phosphorus nanosheets for electrochemical aptasensing of patulin. *Microchim. Acta* **2019**, *186*, 238. [CrossRef]
36. Liu, B.; Abbas, A.; Zhou, C. Two-Dimensional Semiconductors: From Materials Preparation to Electronic Applications. *Adv. Electron. Mater.* **2017**, *3*, 1700045. [CrossRef]
37. Wu, S.; Zhang, H.; Shi, Z.; Duan, N.; Fang, C.; Dai, S.; Wang, Z. Aptamer-based fluorescence biosensor for chloramphenicol determination using upconversion nanoparticles. *Food Control.* **2015**, *50*, 597–604. [CrossRef]
38. Wang, J.; Liu, D.; Huang, H.; Yang, N.; Yu, B.; Wen, M.; Wang, X.; Chu, P.K.; Yu, X.-F. In-Plane Black Phosphorus/Dicobalt Phosphide Heterostructure for Efficient Electrocatalysis. *Angew. Chem.* **2017**, *130*, 2630–2634. [CrossRef]
39. Kang, J.; Wood, J.D.; Wells, S.A.; Lee, J.H.; Liu, X.; Chen, K.S.; Hersam, M.C. Solvent exfoliation of electronic-grade, two-dimensional black phosphorus. *ACS Nano* **2015**, *9*, 3596–3604. [CrossRef]
40. Xia, F.; Wang, H.; Jia, Y. Rediscovering black phosphorus as an anisotropic layered material for optoelectronics and electronics. *Nat. Commun.* **2014**, *5*, 4458. [CrossRef]
41. Wood, J.D.; Wells, S.A.; Jariwala, D.; Chen, K.-S.; Cho, E.; Sangwan, V.K.; Liu, X.; Lauhon, L.J.; Marks, T.J.; Hersam, M.C. Effective Passivation of Exfoliated Black Phosphorus Transistors against Ambient Degradation. *Nano Lett.* **2014**, *14*, 6964–6970. [CrossRef] [PubMed]
42. Goodman, N.B.; Ley, L.; Bullett, D.W. Valence-band structures of phosphorus allotropes. *Phys. Rev. B* **1983**, *27*, 7440–7450. [CrossRef]
43. Brunner, J.; Thüler, M.; Veprek, S.; Wild, R. X-ray photoelectron study of amorphous phosphorus preparedbyplasmachemical transport. Comparison with crystalline polymorphs. *J. Phys. Chem. Solids* **1979**, *40*, 967–971. [CrossRef]
44. Han, Q.; Wang, R.; Xing, B.; Zhang, T.; Khan, M.S.; Wu, D.; Wei, Q. Label-free photoelectrochemical immunoassay for CEA detection based on CdS sensitized WO3@BiOI heterostructure nanocomposite. *Biosens. Bioelectron.* **2018**, *99*, 493–499. [CrossRef] [PubMed]
45. Lai, C.; Zhang, M.; Li, B.; Huang, D.; Zeng, G.; Qin, L.; Liu, X.; Yi, H.; Cheng, M.; Li, L.; et al. Fabrication of CuS/$BiVO_4$ (0 4 0) binary heterojunction photocatalysts with enhanced photocatalytic activity for Ciprofloxacin degradation and mechanism insight. *Chem. Eng. J.* **2018**, *358*, 891–902. [CrossRef]
46. Wu, S.; Duan, N.; Zhang, W.; Zhao, S.; Wang, Z. Screening and development of DNA aptamers as capture probes for colorimetric detection of patulin. *Anal. Biochem.* **2016**, *508*, 58–64. [CrossRef]
47. Bagheri, N.; Khataee, A.; Habibi, B.; Hassanzadeh, J. Mimetic Ag nanoparticle/Zn-based MOF nanocomposite (AgNPs@ ZnMOF) capped with molecularly imprinted polymer for the selective detection of patulin. *Talanta* **2018**, *179*, 710–718. [CrossRef]
48. Malysheva, S.V.; Di Mavungu, J.D.; Boonen, J.; De Spiegeleer, B.; Goryacheva, I.Y.; Vanhaecke, L.; De Saeger, S. Improved positive electrospray ionization of patulin by adduct formation: Usefulness in liquid chromatography–tandem mass spectrometry multi-mycotoxin analysis. *J. Chromatogr. A* **2012**, *1270*, 334–339. [CrossRef]
49. Wu, Z.; Xu, E.; Jin, Z.; Irudayaraj, J. An ultrasensitive aptasensor based on fluorescent resonant energy transfer and exonuclease-assisted target recycling for patulin detection. *Food Chem.* **2018**, *249*, 136–142. [CrossRef]
50. Funari, R.; Della Ventura, B.; Carrieri, R.; Morra, L.; Lahoz, E.; Gesuele, F.; Altucci, C.; Velotta, R. Detection of parathion and patulin by quartz-crystal microbalance functionalized by the photonics immobilization technique. *Biosens. Bioelectron.* **2015**, *67*, 224–229. [CrossRef]

51. Yu, Y.; Fan, Z. Determination of patulin in apple juice using magnetic solid-phase extraction coupled with high-performance liquid chromatography. *Food Addit. Contam. Part A* **2017**, *34*, 273–281. [CrossRef]
52. Zhang, W.; Han, Y.; Chen, X.; Luo, X.; Wang, J.; Yue, T.; Li, Z. Surface molecularly imprinted polymer capped Mn-doped ZnS quantum dots as a phosphorescent nanosensor for detecting patulin in apple juice. *Food Chem.* **2017**, *232*, 145–154. [CrossRef]

foods

MDPI

Article

Au Nanoparticles Functionalized Covalent-Organic-Framework-Based Electrochemical Sensor for Sensitive Detection of Ractopamine

Shuying Yang †, Ruixi Yang †, Jingyi He, Yu Zhang, Yahong Yuan, Tianli Yue and Qinglin Sheng *

Laboratory of Nutritional and Healthy Food-Individuation Manufacturing Engineering, Research Center of Food Safety Risk Assessment and Control, College of Food Science and Technology, Northwest University, Xi'an 710069, China
* Correspondence: qlsheng@nwu.edu.cn
† These authors contributed equally to this work.

Abstract: Ractopamine, as a feed additive, has attracted much attention due to its excessive use, leading to the damage of the human nervous system and physiological function. Therefore, it is of great practical significance to establish a rapid and effective method for the detection of ractopamine in food. Electrochemical sensors served as a promising technique for efficiently sensing food contaminants due to their low cost, sensitive response and simple operation. In this study, an electrochemical sensor for ractopamine detection based on Au nanoparticles functionalized covalent organic frameworks (AuNPs@COFs) was constructed. The AuNPs@COF nanocomposite was synthesized by in situ reduction and was characterized by FTIR spectroscopy, transmission electron microscope and electrochemical methods. The electrochemical sensing performance of AuNPs@COF-modified glassy carbon electrode for ractopamine was investigated using the electrochemical method. The proposed sensor exhibited excellent sensing abilities towards ractopamine and was used for the detection of ractopamine in meat samples. The results showed that this method has high sensitivity and good reliability for the detection of ractopamine. The linear range was 1.2–1600 μmol/L, and the limit of detection (LOD) was 0.12 μmol/L. It is expected that the proposed AuNPs@COF nanocomposites hold great promise for food safety sensing and should be extended for application in other related fields.

Keywords: food safety; food additive sensing; ractopamine; electrochemical detection; covalent organic frameworks; nanocomposites; Au nanoparticles

1. Introduction

In recent years, food safety has become a global concern due to the popularity of healthy diets. Due to the diversity of food types, the unknown source of pollution and the accuracy and timeliness of test results, there are also some regulatory blind spots in the field of food supervision. Therefore, the development of rapid, efficient and multiple food pollutant detection technologies has always been a difficult and hot spot in food safety research. In order to improve the taste of the product, some additives are added to food and livestock products. Among them, as an important neurotransmitter in the human central nervous system, ractopamine [1,2] is added to the feed to make livestock and poultry muscle grow faster, increasing lean meat content. However, if it is not used in a standardized manner, it will cause irreversible damage to the human nervous system and physiological function. Proclamation No. 176 issued by the Ministry of Agriculture in 2010 in China banned the use of ractopamine hydrochloride in feed and animal drinking water. The current Chinese national standard GB/T 20189-2006 is "Determination of Ractopamine in Animal Feed by High Performance Liquid Chromatographic Method", which stipulated that the addition of ractopamine hydrochloride in feed is prohibited. This standard also influences China's imports and exports of the pork trade worldwide [3]. Therefore, it is of great practical value to construct an analysis and detection technology for ractopamine in food samples.

There have been many analytical methods for the determination of ractopamine at present, such as high-performance liquid chromatography [4], liquid chromatography–tandem mass spectrometry [5], gas chromatography–mass spectrometry [6], enzyme-linked immunosorbent assay [7]. However, the above methods are costly, laborious and lack reproducibility. Electrochemical sensing has advantages of low cost, sensitive response and simple operation [8,9], causing serious concern in the fields of pharmacy, clinical diagnosis, environmental monitoring, food safety, etc. [10–12]. Up to now, various new nanomaterials have been used to prepare valuable new electrochemical ractopamine sensors. Electrochemical sensors are considered to be the most effective method for detecting ractopamine [13]. The use of modified electrodes to detect ractopamine is a constant research area in food, clinical and health areas. For examples, Mahmoud et al. reported an electrochemical biodevice developed for the highly selective detection of ractopamine in feed under a double recognition strategy of the aptamer (Apt) and molecular imprinting polymer (MIP) on a glassy carbon electrode (GCE). The detection limit was 330 aM and the linear range was 1 fM to 1.90 μM. In addition, the senor was also applied to urine samples [14]. Cao L.P. et al. developed a kind of electrochemical biosensor using a biochar-supported Cu^{2+}/Cu^{+} composite as an electrochemical sensing interface for detecting ractopamine, which was beneficial to the accumulation and electrochemical sensing of targets. The good linear relationship was in the range of 0.1–1.75 μM, and the detection limit was 0.041 μM [15]. Murugan K et al. prepared a disposable electrochemical sensor using titanium nanoparticles (Ti NPs)-anchored functionalized multi-walled carbon nanotube (Ti@f-MWCNTs) composite as electrochemical sensing interface for the detection of ractopamine. The sensor had a wide detection range (0.01–185 μM) for ractopamine and a low detection limit of 0.0038 μM [16]. Janus particles can also be used as a sustainable electrode modifier material for the electrochemical monitoring of the ractopamine [17]. However, most electrode materials have various defects in terms of a low effective electroactive surface area and lack of predictable structure. Therefore, seeking a new type of composite nanomaterials with controllable structure, high specific surface area and good catalytic performance has become the main trend to improve the electrochemical performance.

Since first being introduced by Yaghi et al. in 2005, COFs have been prepared in various shapes and multi-dimensional frameworks (2D-COFs, 3D-COFs) [18–20]. COFs have been widely used in the field of electrochemical sensing due to their excellent properties [21,22]. Compared with MOF materials, COFs have the advantages of low cost, easy synthesis, low toxicity and better biocompatibility [23]. On the one hand, the ordered microporous or mesoporous structure of COFs can effectively limit the agglomeration and growth of metal nanoparticles during the reduction process, thereby stabilizing the ultrafine nanostructures and regulating the electronic structure and stereochemistry during the catalytic process. On the other hand, COFs have a large specific surface area, providing a large active surface for the loading of electroactive molecules. In addition, their high acid and alkali resistance and excellent biocompatibility can further improve the stability of electrochemical sensors [24]. Tian et al. [25] successfully prepared a novel composite material using COF nanosheets as a matrix to support the growth of gold nanoparticles, after which the synthesized hybrid nanosheets were used as a new platform for multiplexed detection of hepatitis A virus DNA and hepatitis B virus DNA. Moreover, the proposed sensor was also applied for K^{+} sensing and imaging in living cells. Xin et al. [26] used COFs to modify carbon paste electrodes for electrochemical detection of hydroquinone and catechol. The results show that COF-modified carbon material is a promising candidate material for electrochemical detection of phenolic compounds. Functionalization of the synthesized COFs nanomaterials can further improve the selectivity, specificity and sensitivity of functional nanomaterials to target trace substances [27]. Among them, metal nanoparticles, especially AuNPs promote electron transfer at the electrochemical sensing interface as an efficient catalyst based on AuNPs@COFs.

In this work, nanocomposites (AuNPs@COFs) were successfully prepared by in situ reduction method at room temperature using COFs materials as precursors. The prepared nanocomposites were characterized by TEM, FTIR and Zeta potential analyzers to analyze their morphology, element composition and surface charge. The electrochemical sensing performance of AuNPs@COF-modified glassy carbon electrode for ractopamine was investigated by electrochemical method. Based on AuNPs@COF nanocomposites, a ractopamine electrochemical sensor was constructed to detect ractopamine in actual food samples.

2. Materials and Methods

2.1. Chemicals and Reagents

2,5-Divinyl-1,4-phthalaldehyde (DVA, 98%) and 1,3,5-tris (4-aminophenyl) benzene (TAPB, >93.8%) were purchased from Jilin Chinese Academy of Sciences-Yanshen Technology Co., Ltd (Jilin, China). Acetonitrile (C_2H_3N, >99%), acetic acid (AcOH, 99.5%), tetrahydrofuran (THF, 99.9%), sodium citrate ($C_6H_5Na_3O_7$, $\geq$99%), dopamine ($\geq$99.8%), isoxsuprine (>99%), ritodrine (99.5%), fenoterol (99%) and ethanol (C_2H_6O, $\geq$99.7%) were purchased from Shanghai Aladdin Biochemical Technology Co., Ltd. (Shanghai, China) Chloroauric acid ($HAuCl_4$) was purchased from Shanghai Sixin Biotechnology Co., Ltd. (Shanghai, China) Ascorbic acid (AA, $\geq$99.7%) was purchased from Tianjin Tianli Chemical Reagent Co., Ltd. (Tianjin, China) Uric acid (UA, $\geq$99%) and glucose (Glu) were purchased from Tianjin Kermel Chemical Reagent Co., Ltd. (Tianjin, China) Ractopamine was purchased from Energy Chemical Technology Co., Ltd. (Shanghai, China).

2.2. Apparatus and Characterization

A G2 F20 S-TWIIN high-resolution transmission electron microscope was used in this experiment (FEI Tennessee, USA). A Fourier transform infrared spectrometer (FT-IR) was recorded using a Vertrx 70 FT-IR spectrometer (Broker, Germany). The Zeta potential were measured using a Zeta potential analyzer (Malvern Panalytical, UK). All electrochemical and current-voltage measurements were performed on a CHI660D electrochemical analyzer (Shanghai Chenhua Instrument Co., Ltd., Shanghai, China).

2.3. Preparation of COFs Composite Nanomaterials

Spherical COFs nanomaterials were prepared by the following steps [28]:

Preparation and optimization of COFs: The uniform morphology spherical COFs nanomaterials were synthesized under the action of catalyst between amino groups and aldehyde groups by polycondensation at room temperature and then simply cleaned. Firstly, 0.014 g of TAPB and 0.011 g of DVA were put together in a 10 mL centrifuge tube. Then, 5 mL acetonitrile was added to the centrifuge tube and was treated with ultrasound for 2 min. Then, 0.2, 0.4, 0.6, 0.8, 1.0, 1.2, 1.4, 1.6 mL, 12 mol/L glacial acetic acid were dripped, respectively. Subsequently, the mixture was shaken vigorously for 15 s and allowed to stand at room temperature for 72 h. The obtained yellow precipitate was centrifuged at 9600 rpm, and then washed several times with anhydrous THF and ethanol until the lotion was clear. Finally, the resulting sample was dried at a high vacuum of 60 °C for about 24 h.

Preparation of AuNPs@COFs nanocomposites: AuNPs@COFs nanocomposites were prepared by in situ growth method at room temperature [29]. The 7 mg spherical COFs with the best morphology and size were transferred to 10 mL of 1 mmol/L $HAuCl_4$ solution. The mixture was then stirred on a magnetic stirrer for 5 min. Subsequently, the sodium citrate solution was slowly added. After 30 min of reaction at room temperature, the mixture was centrifuged at 6000 rpm for 5 min. To collect the pure composite material, the precipitate was washed four times with ultrapure water. Finally, the obtained dark-brown product was dried at 60 °C in vacuum for 16 h.

2.4. Construction of AuNPs@COFs Nanocomposites Electrochemical Sensor

Before modification, the bare glassy carbon electrode (GCE) with a diameter of 3 mm was ground with 0.05 μm Al_2O_3 powder, washed with ultrapure water, and then ultrasonically cleaned with a mixture of ethanol and ultrapure water for 2 min. After drying, the surface of the bare GCE remained as smooth as a mirror. 1 mg AuNPs@COFs nanocomposites were uniformly dispersed in 1 mL 0.5% chitosan solution to obtain a mixed solution with a concentration of 1 mg/mL. Finally, 7 μL of the mixed solution was dropped on the surface of the prepared GCE and dried naturally at room temperature to obtain the modified electrode AuNPs@COFs/chitosan/GCE. At the same time, COFs/chitosan/GCE were prepared using the same method. Figure 1 shows the electrode modification and subsequent detection process.

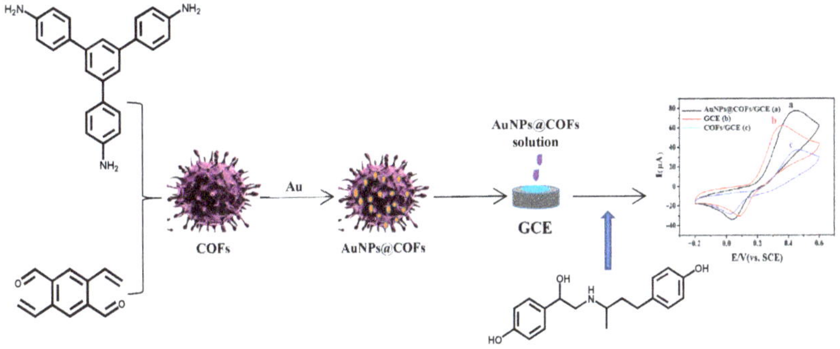

Figure 1. Preparation and sensing principle diagram of AuNPs@COFs nanocmposites.

2.5. Pretreatment of Actual Samples

Pork and chicken were purchased at the local market. The two samples were processed in the following steps [30]: 2.0 g of pork and chicken samples were weighed, added in acetonitrile, and were twisted into a uniform paste. Then, the homogeneous solution was centrifuged at 6000 rpm for 9 min, and the supernatant was collected. Next, the precipitate was dissolved in acetonitrile, and the above steps were repeated again. Secondly, the supernatant was collected. The twice-collected supernatant was mixed and filtered through a membrane with a pore size of 0.2 μm. The collected liquid was evaporated to 2 mL. Finally, the resulting extract was precisely added into ultrapure water to 5 mL and stored in a refrigerator at 4 °C.

3. Results and Discussion

3.1. Characterization of COFs Nanomaterials and AuNPs@COFs Nanocomposites

As shown in Figure 2, the prepared COFs nanomaterials have a certain uniformity and dispersion. When the concentration of acetic acid (HAc) remained unchanged at 12 mol/L, the average diameter of spherical COFs increased from 250 nm to 1.1 μm while the amount of HAc increased from 0.2 mL to 1.6 mL. The effect of HAc content on the morphology of COFs was also investigated. In Figure 2A–H, the amounts of HAc are 0.2, 0.4, 0.6, 0.8, 1.0, 1.2, 1.4, 1.6 mL. The particle diameter distributions of the corresponding COFs were shown in Figure 2B–G. Respectively, the average particle sizes of the COFs are ~1.1 μm (Figure 2A), ~800 nm (Figure 2B), ~700 nm (Figure 2C), ~400 nm (Figure 2D), ~350 nm (Figure 2E), ~300 nm (Figure 2F), ~270 nm (Figure 2G), ~250 nm (Figure 2H). It can be clearly seen from the figure that with the increase in the amount of HAc, the size of the COF nanomaterials gradually decreased. At the same time, the morphology also changed significantly. The more HAc that was used, the more burrs appeared on the outer layer of COFs, changing from the initial sphere to gradually become more flower-like. As the particle size decreased, the specific surface area increased, and the outer surface became rough. This not only creates a favorable environment for the attachment of AuNPs, but also lays the foundation for the construction of electrochemical sensing experiments based on AuNPs@COF nanocomposites.

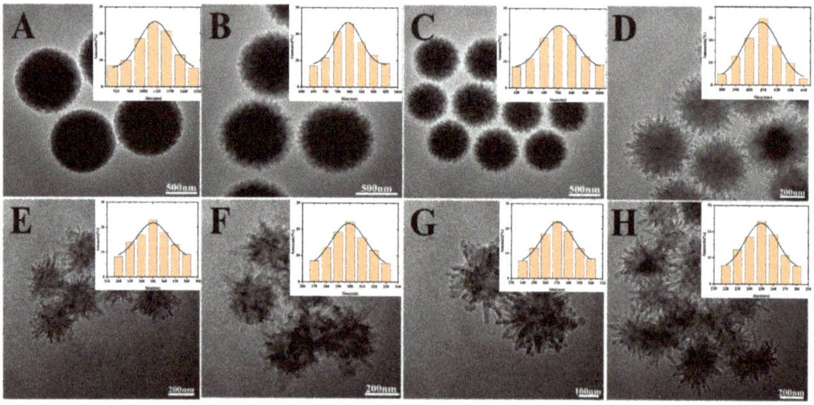

Figure 2. TEM images and particle diameter distribution of uniform spherical COFs nanomaterials with different sizes prepared in 0.2, 0.4, 0.6, 0.8, 1.0, 1.2, 1.4 and 1.6 mL HAc solution: (**A**) 0.2 mL, 1100 nm; (**B**) 0.4 mL, 800 nm; (**C**) 0.6 mL, 700 nm; (**D**) 0.8 mL, 400 nm; (**E**) 1 mL, 350 nm; (**F**) 1.2 mL, 300 nm; (**G**) 1.4 mL. 270 nm; (**H**) 1.6 mL, 250 nm.

Firstly, the optimized COFs nanomaterials were screened, and then the selected COFs nanomaterials with the best morphology and size were functionalized with AuNPs. The prepared COFs nanomaterials and AuNPs@COFs were characterized by TEM. The results are shown in Figure 3A,B. The COF nanomaterials prepared in Figure 3A are uniform in size and neatly arranged. The selected material was prepared by 0.8 mL HAc with a particle diameter distribution in the range of 370–430 nm and an average particle size of about 400 nm, as shown in Figure 3C. Relatively, the particle diameter distribution range of AuNPs@COFs nanocomposites was 380–440 nm and the average particle size was 410 nm (Figure 3D). The nanostructure is similar to a sea urchin, which is caused by the polycondensation reaction during the catalytic process. This structure is conducive to ion transfer during the electrochemical catalytic process. At the same time, the larger specific surface area of COFs nanomaterials provides more loading sites, and can also prevent AuNPs from agglomeration during growth to some extent. In Figure 3B, the flower-like nanostructure of AuNPs@COF was prepared using the in situ reduction method under the action of sodium citrate. The black small particles attached in Figure 3B are dispersed AuNPs with the particle size of 7 nm. The successful attachment of AuNPs greatly improves the electrical conductivity of the COFs material and allows for a large amount of space inside and outside the structure, which is beneficial to electron transfer and conduction.

The functional groups formed during the preparation of the material were analyzed by FTIR [28]. Figure 4A (a) shows that a characteristic peak belonging to C=N appears in COFs at 1606 cm^{-1}, which indicates that the aldehyde group of DVA and the amino group of TAPB undergo condensation reaction under the action of catalyst. As can be seen from Figure 4A (b), after the introduction of AuNPs on the surface of the COFs, the presence of the characteristic peak C=N at 1606 cm^{-1} is the same as that of COFs, further indicating that the structure of COFs is still intact and not affected by AuNPs doping. In addition, it is worth noting that the observed absorption band shifted from 2360 cm^{-1} to 2363 cm^{-1}, which may be attributed to the electrostatic interaction between NH_3^+ and $[AuCl_4]^-$ [31]. Thus, the AuNPs@COFs nanocomposites were successfully prepared. Zeta potential analyzer was used to study the charge on the surface of nanomaterials before and after composite. Figure 4B, (a) shows that the charge on the surface of COFs material was −27.86 mV, and (b) shows that the charge on the surface of AuNPs@COFs nanocomposites was −10.28 mV. Due to the positive charge of AuNPs, the negative charge on the surface of the nanocomposites slightly reduced. In summary, the surface of COF nanomaterials was negatively charged before and after functionalization.

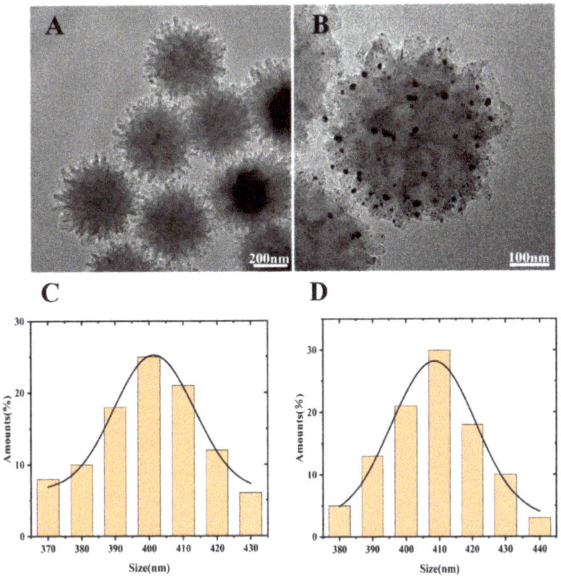

Figure 3. TEM images of COF nanomaterials (**A**) and AuNPs@COF nanocomposites (**B**); COF nanomaterials (**C**); and AuNPs@COFs nanocomposites' (**D**) particle diameter distribution.

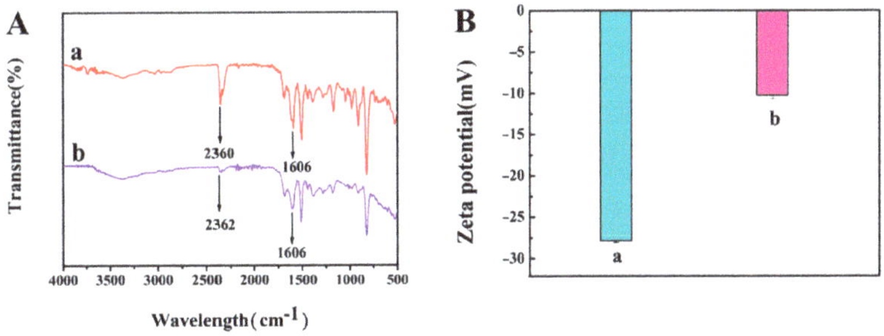

Figure 4. (**A**) FTIR spectra of COFs nanomaterials (a) and AuNPs@COFs nanocomposites (b); (**B**) zeta plots of COFs nanomaterials (a) and AuNPs@COFs nanocomposites (b).

3.2. Electrochemical Performance of AuNPs@COFs Nanocomposites

The electrochemical properties of different modified electrodes in $[Fe(CN)_6]^{3-/4-}$ solution were studied by cyclic voltammetry (CV). Figure 5A shows the CV curves of different modified electrodes with a pair of well defined redox peaks, characteristic of the $[Fe(CN)_6]^{3-/4-}$ redox processes. The redox current of COFs/GCE(c) electrode was lower than that of the bare GCE(b) electrode, which may be due to the poor conductivity of COFs. The redox current of COFs/GCE(c) electrode was much lower than that of the AuNPs@COFs/GCE(a) electrode. The enhancement of redox current of the AuNPs@COFs/GCE(a) electrode may be due to the direct band gap and high electron transfer ability of AuNPs, which brought the AuNPs@COFs excellent conductivity, strong catalytic activity and a large specific surface area. The results show that AuNPs@COFs exhibit the strongest peak current response, indicating that the synergistic effect of the nanocomposites improves the electron transfer rate and catalytic activity. The heterogeneous electron transfer rate (k^0) was also calculated using the method developed by Klingler and Kochi [32]. The results showed that AuNPs@COFs/GCE exhibited faster transfer kinetics with $k^0 = 2.74 \times 10^{-3}$ cm s^{-1}, which was faster than that of COFs/GCE ($k^0 = 1.08 \times 10^{-3}$ cm s^{-1}). The results show that AuNPs@COFs have good levels of conductivity and are suitable as electrode modification materials for electrochemical sensors.

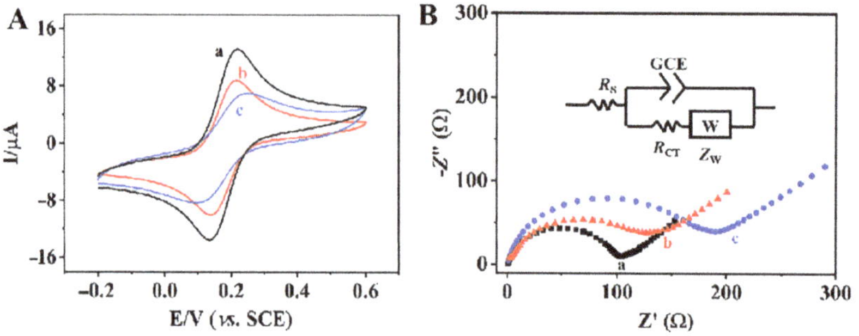

Figure 5. (**A**) CV curves of AuNPs@COFs (a), GCE (b) and COFs (c) in in 0.5 mmol/L $[Fe(CN)_6]^{3-/4-}$ and 0.1 mol/L KCl electrolyte solution; (**B**) EIS of AuNPs@COFs nanocomposites (a), GCE (b) and COFs (c) in 0.5 mmol/L $[Fe(CN)_6]^{3-/4-}$ and 0.1 mol/L KCl electrolyte solution.

In order to study the electrochemical properties of different materials modified electrodes, the interface properties of different electrodes were evaluated by electrochemical impedance spectroscopy (EIS) in the frequency range of 0.01–100 kHz. Figure 5B shows the Nyquist plots of different modified electrodes, where the semicircle diameter in the high frequency region represents the electron transfer resistance and the linear portion in the low frequency region represents the diffusion resistance [33]. Nyquist plots (Figure 5B) consisting of an imaginary part (Z'') versus a real part (Z') were simulated according to the Randles equivalent circuit model shown in Figure 5B—inset. The circuit consists of the R_S component, which is the ohmic resistance from the solution, in series with the combination of constant phase element (GCE) combined in parallel with the charge transfer resistance (R_{CT}) of the Faradaic process and the Wargburg impedance (Z_W) component arising from the mass diffusion towards the electrode surface. Herein, the GCE component is incorporated instead of the true double layer capacitance (C_{dl}) due to heterogeneity in the electrode surface that contributes to the imperfect capacitance. The Rct value of COFs (c) was about 170 Ω and the conductivity was low. The Rct value of the bare GCE (b) was about 130 Ω, whereas the Rct value of AuNPs@COFs (a) decreased to about 90 Ω, which is due to the good conductivity of AuNPs and almost no agglomeration on the surface of COFs. It shows that different modified components have obvious synergistic effects. This result is consistent with the above CV curve. In summary, AuNPs@COF nanocomposites exhibit good charge transfer kinetics and a fast electron transfer.

The response of different modified electrodes to ractopamine was studied by cyclic voltammetry (CV). As shown in Figure 6, the peak current response of COFs/GCE (c) to ractopamine was lower than that of bare GCE (b), which may be due to the poor conductivity of COF, and the introduction of AuNPs significantly improved the peak current response of COFs/GCE. AuNPs@COFs/GCE (a) showed the strongest peak current response to ractopamine, indicating that the composite material improved the electron transfer rate and catalytic activity; therefore, AuNPs@COFs/GCE was selected for ractopamine detection.

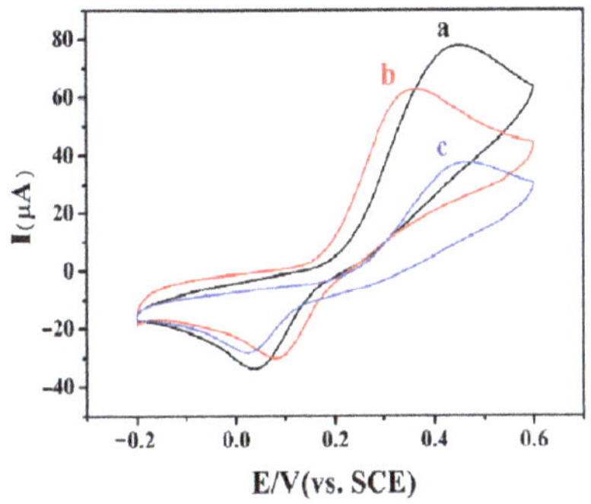

Figure 6. CV curves of AuNPs@COFs (a), GCE (b) and COFs (c) in 0.1 mol/L PBS electrolyte containing 0.6 mmol/L ractopamine.

In order to study the electrode reaction kinetics, the effect of scanning rate on ractopamine current and potential was studied using the CV method. In Figure 7A, when the concentration of ractopamine was 1 mmol/L and the scan rate increased from 20 mV/s to 200 mV/s, the redox peak current of ractopamine increased linearly. As shown in Figure 7B, the anode and cathode peak currents of ractopamine are fitted to the square root of the scanning speed ($\upsilon^{1/2}$). The regression equation is as follows:

$$Ipa = 5.15\upsilon^{1/2}(mV/s)^{1/2} - 0.3427 \ (R^2 = 0.9997) \quad (1)$$

$$Ipc = -1.3355\upsilon^{1/2}(mV/s)^{1/2} + 1.5367 \ (R^2 = 0.9988) \quad (2)$$

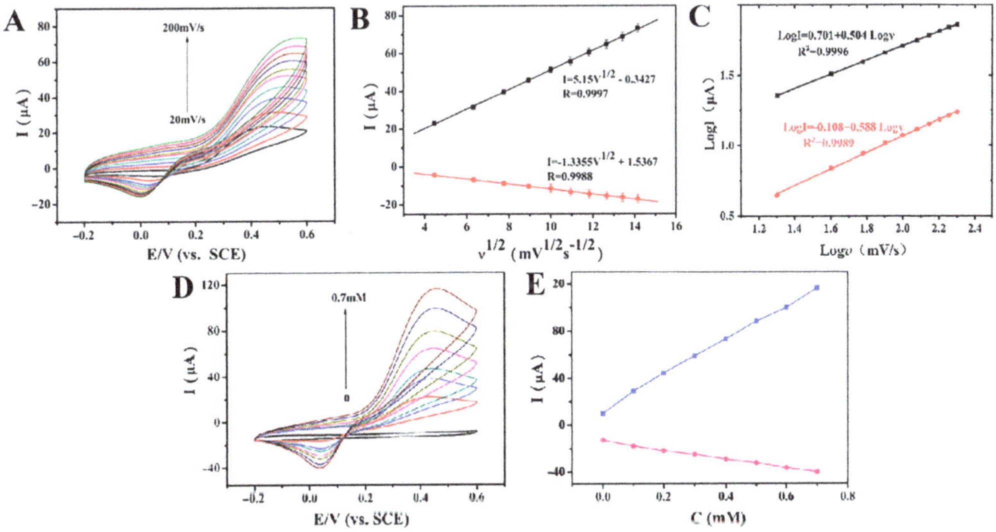

Figure 7. CV curves of AuNPs@COFs nanocomposites to 1 mmol/L ractopamine at different scan rates (20, 40, 60, 80, 100, 120, 140, 160, 180, 200 mV/s) in 0.1 mol/L PBS (**A**) and the relationship between peak current and square root of scan rate (**B**); corresponding logI vs. log$v$ plots (**C**); CV curves of AuNPs@COFs nanocomposites for different concentrations of ractopamine (0, 0.1, 0.2, 0.3, 0.4, 0.5, 0.6 and 0.7 mmol/L) (**D**); linear relationship between ractopamine concentration and ractopamine peak current (**E**).

The oxidation peak current and reduction peak current are proportional to the square root of the scan rate, which is consistent with the quantitative relationship between Ip and $\upsilon^{1/2}$ in the Randles–Sevcik equation, indicating that the electron loss process on the surface of the AuNPs@COFs composites is controlled by diffusion [34]. It shows that the material has high catalytic performance under ractopamine. Moreover, with the increase in the scanning rate, the anodic peak potential (Epa) of ractopamine shifts positively and the cathodic peak potential (Epc) shifts negatively. This is due to the kinetic control of the redox reaction process. The relationship curves between logI and logυ were also studied. From Figure 7C, it could be seen that the slopes of Figure 7B curves corresponding to logI and logυ were 0.504 and 0.588, respectively, reflecting the fast kinetics dominated by pseudocapacitance process.

In order to further investigate the electrochemical catalytic activity of AuNPs@COFs nanocomposites for ractopamine, CV analysis was performed at a fixed scanning rate (100 mV/s) and a certain ractopamine concentration range (0, 0.1, 0.2, 0.3, 0.4, 0.5, 0.6, 0.7 mmol/L) (Figure 7C). The results showed that with the increase in ractopamine concentration, the oxidation peak current of ractopamine gradually increased and the amplitude was larger, whereas the reduction peak current increased less, indicating that the oxidation peak current was more sensitive to the determination of ractopamine. Figure 7D,E show that the concentration of ractopamine is positively correlated with the oxidation peak and reduction peak current of ractopamine.

3.3. *Amperometric Detection of Ractopamine*

It is necessary to optimize the working potential before quantitative analysis of ractopamine. The working potential affects the sensitivity of the sensor. Figure 8 is the I-T diagram of AuNPs@COFs nanocomposites measured under the condition of changing the concentration of ractopamine (adding 10 μL 0.1 mol/L ractopamine every 50 s) by continuously and stably stirring after changing different working potentials. It was found that the current reached the maximum voltage at 0.40 V when the working potential increased from 0.30 V to 0.45 V. Therefore, 0.40 V was selected as the optimal potential and used for subsequent quantitative analysis.

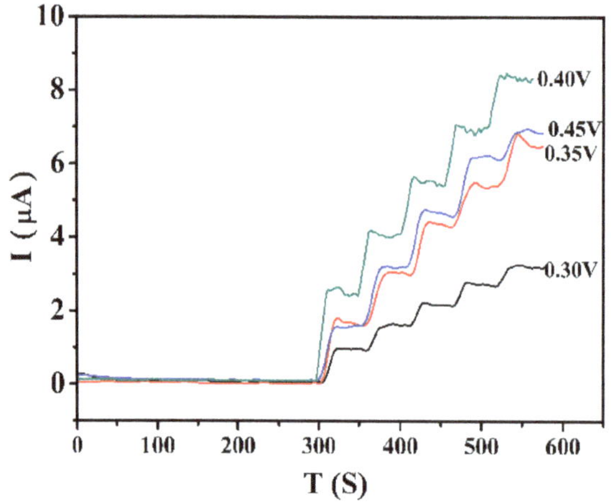

Figure 8. I-T plots of AuNPs@COFs/GCE to ractopamine at different potentials in 0.1 mol/L PBS solution.

According to the above method, ractopamine was added to 10 mL PBS electrolyte every 50 s. Figure 9A shows the process of current changing with time. With the continuous addition of ractopamine, the current showed a stepwise upward trend. There are two corresponding current and concentration calibration diagrams for the oxidation of ractopamine by AuNPs@COFs/GCE in different ractopamine concentration ranges. The linear relationship is I (μA) = 0.08C (μmol/L) + 5.55 in the range of 1.2–280 μmol/L, and I (μA) = 0.03C (μmol/L) + 18.73 in the range of 280–1600 μmol/L (Figure 9B). R^2 is 0.96 and 0.98, respectively. When the ractopamine concentration was high (>3.1 mmol/L), the current was nonlinear, which may be ascribed to the saturation of the catalytic sites on the working electrode surface. The calculated LOD [35] is 0.12 μmol/L. In order to confirm the accuracy and precision of this method, we also performed F test and *t* test. The t values of the two linear ranges were 12.81 and 14.53, and the F values were 1918.98 and 932.27, $p < 0.01$, respectively, which shows that the method has a significant linear relationship, and good accuracy and precision. Compared with other reported ractopamine sensors (Table 1), AuNPs@COFs/GCE has a wider linear range, which indicates that AuNPs@COFs nanocomposites are good catalysts for ractopamine.

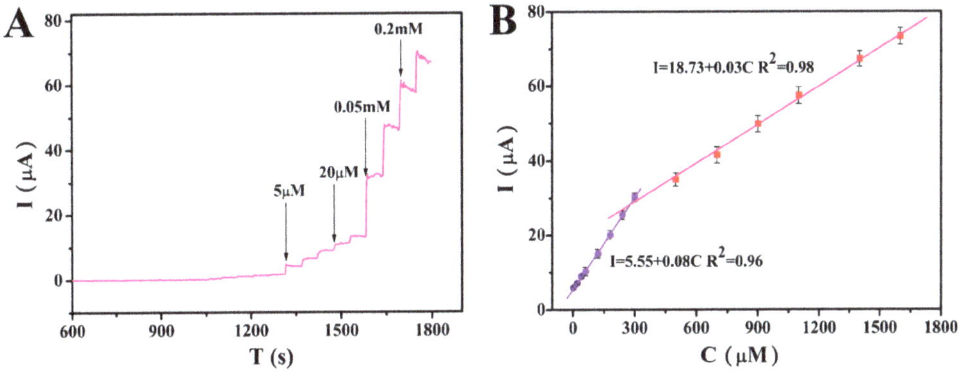

Figure 9. (**A**) The I-T diagram of AuNPs@COFs/GCE for ractopamine was obtained by adding different concentrations of ractopamine into 0.1 mol/L PBS solution at 0.40 V applied voltage. (**B**) Linear fitting curve of ractopamine concentration and current value.

Table 1. Performance comparison of different modified electrodes for ractopamine electrocatalytic oxidation.

Electrode *	Linear Range (μmol/L)	Limit of Detection (LOD) (μmol/L)	Reference
SMWCNT-NF/ITO	0.3–33.3	0.1	[35]
PoAT-AuNP-Au/SCE	2.5–150	1.17	[36]
FLG-PI/FGE	5–80	1.29	[37]
	10–200	2.7	
	25–250	7.81	
$NiFe_2O_4$/GCE	2–60	1	[38]
Nafion/HPMo/MoS_2/PDDA/GCE	1–70	0.06	[39]
AuNPs@COFs/GCE	1.20–280280–1600	0.12	This work

* SMWCNT-NF: Single-walled carbon nanotubes-nafion; PoAT: Poly(o-aminothiophenol); FLG-PI: Layers of graphene-polyimide film; HPMo: Polyoxometalate $H_3PMo_{12}O_{40}$. PDDA: Polydiene dimethyl ammonium chloride solution.

3.4. Selectivity, Stability and Repeatability Analysis of Proposed Sensor

Anti-interference ability is one of the most important indexes to test the practicability of electrode. AuNPs@COFs nanocomposites were used to detect some electroactive substances coexisting with glucose in real samples. Potential interferences such as hydrogen peroxide (H_2O_2), ascorbic acid (AA), glucose (Glu), uric acid (UA), dopamine, isoxsuprine, ritodrine, and fenoterol were detected individually using amperometric assay. As shown in Figure 10A, when 3 mmol/L dopamine, isoxsuprine, 3 mmol/L ritodrine, 3 mmol/L fenoterol, 3 mmol/L Glu, 3 mmol/L H_2O_2, 3 mmol/L UA, and 3 mmol/L AA were added to the system in turn, the response current value remained basically unchanged. However, significant responses were obtained after adding ractopamine. These indicate that AuNPs@COFs/GCE has good anti-interference ability with ractopamine. In addition, six AuNPs@COFs/GCE electrodes were prepared using the same method to test the reproducibility. The relative standard deviation (RSD) was 0.51%, and the maximum increase of the current response was 5.31%, indicating that AuNPs@COFs/GCE have good reproducibility. It can be seen from Figure 10B that the prepared electrode was stored at 4 °C for 15 days to test its long-term stability. The results show that the response current maintained almost 96.4% of its original current, indicating that the sensor based on AuNPs@COF nanocomposites has good stability.

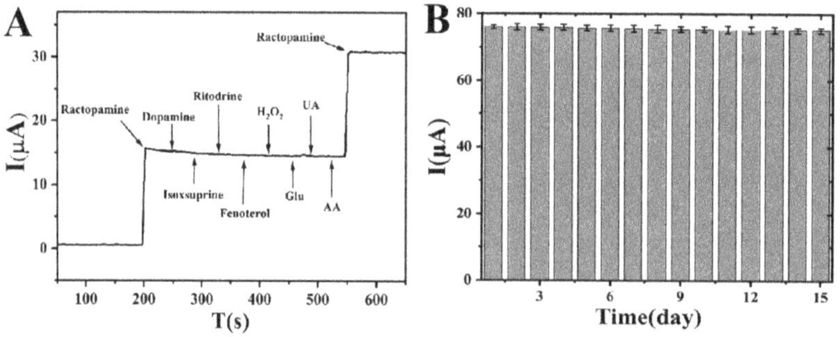

Figure 10. I-T diagram (**A**) in response of AuNPs@COFs electrode in 0.1 mol/L PBS solution when ractopamine, dopamine, isoxsuprine, ritodrine, fenoterol, H_2O_2, Glu, UA, AA and ractopamine were added in turn (the concentration of ractopamine added was 0.3 mmol/L, and the concentration of interfering substances added was 3 mmol/L); stability of the electrochemical sensor based on AuNPs@COF nanocomposites (**B**).

3.5. Detection in Actual Samples

The applicability of electrochemical sensors based on AuNPs@COFs nanocomposites was evaluated. Pork and chicken were selected for further detection in this study. The ractopamine in actual samples was analyzed by standard addition method and compared with HPLC. It is found that the results of HPLC and AuNPs@COFs/GCE for the detection of ractopamine content in meat were basically the same (Table 2). The recovery rate of ractopamine by the sensor was 95.0–108.1% and RSD was 2.13–5.12%. All actual sample detection was performed in meat extracts. This indicates that the sensor can be applied to the detection of ractopamine in actual samples.

Table 2. Recovery and relative standard deviation (RSD, %, n = 3) of ractopamine in meat by electrochemical sensor and HPLC, respectively.

Sample	Ractopamine Added (mmol/L)	HPLC Determination Concentration (mmol/L)	Electrochemical Sensor		
			Determination Concentration (mmol/L)	Relative Standard Deviation (RSD)	Recovery Rate
Sample 1	0.05	0.08	0.08	2.51%	98.0%
	1.00	1.06	0.94	5.12%	95.2%
	2.00	2.10	2.08	3.13%	101.0%
Sample 2	0.1	0.16	0.15	4.01%	97.5%
	1.00	1.10	1.07	3.73%	98.3%
	2.00	2.09	2.11	2.13%	108.1%

4. Conclusions

In summary, COFs have potential expansion value in the field of food safety detection due to their simple synthesis method, stable structure and environmental friendliness. Metal nanoparticles, especially AuNPs, as an efficient catalyst, promote electron transfer on the AuNPs@COF-based electrochemical sensor, overcoming the defects of single metal nanoparticles while combining the advantages of magnetic nanomaterials. In this paper, COFs nanomaterials were synthesized by polycondensation reaction at room temperature, and then AuNPs@COF nanocomposites were prepared by modifying AuNPs on the surface. Subsequently, the nanocomposites were modified on the electrode surface to successfully construct a AuNPs@COFs/chitosan/GCE for ractopamine detection. The results showed that the linear range of ractopamine detection was 1.2–1600 μmol/L and the limit of detection (LOD) was 0.12 μmol/L. Compared with the same type of ractopamine sensor, the sensor based on AuNPs@COF nanocomposites has a wider linear range, lower detection limit, higher stability and better selectivity. The electrochemical sensing platform has been successfully applied to the determination of ractopamine in meat extracts with good accuracy and reliability. Food safety detection methods based on functional nanomaterials will usher in new development opportunities in future.

Author Contributions: Investigation, data curation, formal analysis, methodology, writing, original draft, S.Y. and R.Y.; software, formal analysis, discussion, J.H., Y.Z. and Y.Y.; resources, supervision, review, editing, validation, Q.S. and T.Y. All authors have read and agreed to the published version of the manuscript.

Funding: The National Key R&D Program of China (2019YFC1606703), and the Natural Science Foundation of Shaanxi Province in China (2020JM-429).

Data Availability Statement: The data presented in this study are available on request from the corresponding author.

Acknowledgments: The authors gratefully acknowledge financial support from the National Key R&D Program of China (2019YFC1606703), and the Natural Science Foundation of Shaanxi Province in China (2020JM-429).

Conflicts of Interest: The authors declare no conflict of interest.

References

1. Jing, H.; Ouyang, H.Y.; Li, W.F.; Long, Y. Molten salt synthesis of BCNO nanosheets for the electrochemical detection of clenbuterol. *Microchem. J.* **2022**, *178*, 107359. [CrossRef]
2. Soares, M.H.; Junior, D.T.V.; de Amorim, R.G.; Júnior, R.L.C.; Rocha, G.C.; Bohrer, B.M.; Juárez, M.; de Souza Duarte, M.; Saraiva, A. Effects of feeding ractopamine hydrochloride with or without supplemental betaine on live performance, carcass and meat quality traits, and gene expression of finishing pigs. *Meat Sci.* **2022**, *191*, 108851. [CrossRef] [PubMed]
3. Liu, F. The impact of the Codex Alimentarius Commission ractopamine standard on China's pork trade. *Chin. J. Anim. Husb.* **2015**, *51*, 37–40.
4. Du, W.; Zhao, G.; Fu, Q.; Sun, M.; Zhou, H.; Chang, C. Combined microextraction by packed sorbent and high-performance liquid chromatography-utraviolet detection for rapid analysis of ractopamine in porcine muscle and urine samples. *Food Chem.* **2014**, *145*, 789–795. [CrossRef] [PubMed]
5. Decheng, S.; Peilong, W.; Yang, L.; Ruiguo, W.; Xia, F.; Xiaoou, S. Evaluation of β-agonists in blood meal:Validation of determination method and potential pathway for reentry into the environment. *J. Chromatogr. A* **2019**, *1612*, 460624. [CrossRef] [PubMed]
6. Zhao, M.E.; Peng, Z.C. Comparison and application of three extraction methods in detection of receptor agonist residues in pig liver. *Modern Food* **2019**, *3*, 182–184.
7. *SN/T 4818-2017*; Determination of Ractopamine, Salbutamol and Clenbuterol Hydrochloride in Import and Export Edible Animals Enzyme-Linked Immunosorbent Assay. Chinese Standard. AQSIQ: Beijing, China, 2017.
8. Zhang, Y.; Wang, F.X.; Fang, L.; Wang, S.; Fang, G. Rapid determination of ractopamine residues in edible animal products by enzyme-linked immunosorbent assay: Development and investigation of matrix effects. *Biomed Res. Int.* **2014**, *2009*, 269–281. [CrossRef] [PubMed]
9. Tefera, M.; Tessema, M.; Admassie, S.; Wubet, W. Voltammetric determination of uric acid using multiwall carbon nanotubes coated-poly (4-amino-3-hydroxy naphthalene sulfonic acid) modified glassy carbon electrode. *Heliyon* **2021**, *7*, e07575. [CrossRef]

10. Yan, X.; Song, Y.; Liu, J.; Zhou, N.; Zhang, C.; He, L.; Zhang, Z.; Liu, Z. Two-dimensional porphyrin-based covalent organic framework: A novel platform for sensitive epidermal growth factor receptor and living cancer cell detection. *Biosens. Bioelectron.* **2019**, *126*, 734–742. [CrossRef]
11. Yang, Y.X.; Shen, Y.; Wang, L.; Song, Y.; Wang, L. Three-dimensional porous carbon/covalent-organic framework films integrated electrode for electrochemical sensors. *J. Electroanal. Chem.* **2019**, *855*, 113590. [CrossRef]
12. Liang, H.; Xu, H.T.; Zhao, Y.T.; Zheng, J.; Zhao, H.; Li, G.; Li, C.P. Ultrasensitive electrochemical sensor for prostate specific antigen detection with a phosphorene platform and magnetic covalent organic framework signal amplifier. *Biosens. Bioelectron.* **2019**, *144*, 111691. [CrossRef] [PubMed]
13. Kordasht, H.K.; Saadati, A.; Hasanzadeh, M. A flexible paper based electrochemical portable biosensor towards recognition of ractopamine as animal feed additive: Low cost diagnostic tool towards food analysis using aptasensor technology. *Food Chem.* **2022**, *373*, 131411. [CrossRef] [PubMed]
14. Mahmoud, R.; Mahsa, G.; Faezeh, S.F. Fabrication of an electrochemical biodevice for ractopamine detection under a strategy of a double recognition of the aptamer/molecular imprinting polymer. *Bioelectrochemistry* **2021**, *138*, 107722.
15. Cao, L.P.; Ding, Q.; Liu, M.H.; Lin, H.; Yang, D.P. Biochar-supported Cu^{2+}/Cu^{+} composite as an electrochemical ultrasensitive interface for ractopamine detection. *ACS Appl. Bio Mater.* **2021**, *4*, 1424–1431. [CrossRef]
16. Murugan, K.; Asit, K.P.; Wang, Y.H.; Liu, X.; He, J.H.; Chung, R.J. Titanium nanoparticle anchored functionalized MWCNTs for electrochemical detection of ractopamine in porcine samples with ultrahigh sensitivity. *Food Chem.* **2022**, *378*, 132083.
17. Zhou, Y.; Wang, T.T.; Yan, L.T. A sensitive electrochemical analysis method of ractopamine based on Janus particles. *Chin. J. Anal. Chem.* **2022**, *50*, 1872–2040. [CrossRef]
18. Jin, E.; Geng, K.Y.; Lee, K.H.; Jiang, W.; Li, J.; Jiang, Q.; Irle, S.; Jiang, D. Topology-templated synthesis of crystalline porous covalent organic frameworks. *Angew. Chem.* **2020**, *132*, 12260–12267. [CrossRef]
19. Tan, K.T.; Tao, S.; Huang, N.; Jiang, D. Water cluster in hydrophobic crystalline porous covalent organic frameworks. *Nat. Commun.* **2021**, *12*, 6747. [CrossRef]
20. Liu, X.L.; Li, J.; Gui, B.; Lin, G.; Fu, Q.; Yin, S.; Liu, X.; Sun, J.; Wang, C. A crystalline three-dimensional covalent organic framework with flexible building blocks. *J. Am. Chem. Soc.* **2021**, *143*, 2123–2129. [CrossRef]
21. Zhao, H.; Xie, W.; Zhang, R.L.; Wang, X.D.; Liu, H.F.; Li, J.; Sha, T.; Guo, X.S.; Sun, Q.M.; Zhang, Y.P.; et al. Electrochemical sensor for human norovirus based on covalent organic framework/pillararene heterosupramolecular nanocomposites. *Talanta* **2022**, *237*, 122896. [CrossRef]
22. Sun, Y.; He, J.; Waterhouse, G.I.N.; Xu, L.; Zhang, H.; Qiao, X.; Xu, Z. A selective molecularly imprinted electrochemical sensor with GO@COF signal amplification for the simultaneous determination of sulfadiazine and acetaminophen. *Sens. Actuators B Chem.* **2019**, *300*, 126993. [CrossRef]
23. Wang, J.; Feng, Y.Q.; Zhang, B. MOF-COF hybrid frameworks materials. *Prog. Chem.* **2022**, *34*, 1308–1320.
24. Li, Y.; Guo, X.H.; Li, X.F.; Zhang, M.; Jia, Z.; Deng, Y.; Tian, Y.; Li, S.; Ma, L. Redox-active two-dimensional covalent organic frameworks (COFs) for selective reductive separation of valence-variable, redox-sensitive and long-lived radionuclides. *Angew. Chem.* **2020**, *132*, 4197–4204. [CrossRef]
25. Tian, Y.Y.; Lu, Q.P.; Guo, X.X.; Wang, S.Y.; Gao, Y.; Wang, L.H. Au nanoparticles deposited on ultrathin two-dimensional covalent organic framework nanosheets for in vitro and intracellular sensing. *Nanoscale* **2020**, *12*, 7776–7781. [CrossRef]
26. Xin, Y.; Wang, N.; Wang, C.; Gao, W.; Chen, M.; Liu, N.; Duan, J.; Hou, B. Electrochemical detection of hydroquinone and catechol with covalent organic framework modified carbon paste electrode. *J. Electroanal. Chem.* **2020**, *877*, 114530. [CrossRef]
27. Jin, F.Y.; Nguyen, H.L.; Zhong, Z.Y.; Han, X.; Zhu, C.; Pei, X.; Ma, Y.; Yaghi, O.M. Entanglement of square nets in covalent organic frameworks. *J. Am. Chem. Soc.* **2022**, *144*, 1539–1544. [CrossRef]
28. Ma, W.; Zheng, Q.; He, Y.; Li, G.; Guo, W.; Lin, Z.; Zhang, L. Size-controllable synthesis of uniform spherical covalent organic frameworks at room temperature for highly efficient and selective enrichment of hydrophobic peptides. *J. Am. Chem. Soc.* **2019**, *141*, 18271–18277. [CrossRef]
29. Li, W.; Li, Y.; Qian, H.L.; Zhao, X.; Yang, C.X.; Yan, X.P. Fabrication of a covalent organic framework and its gold nanoparticle hybrids as stable mimetic peroxidase for sensitive and selective colorimetric detection of mercury in water samples. *Talanta* **2019**, *204*, 224–228. [CrossRef]
30. Zhang, X.L.; Luo, Y.J.; Jiang, J.; Lu, Y.; Feng, N. Determination of five β2-agonist residues in pork by ultra-high performance liquid chromatography-tandem mass spectrometry with molecularly imprinted solid phase extraction. *Chin. J. Chromatogr.* **2015**, *33*, 838. [CrossRef]
31. Zhang, T.; Chen, Y.; Huang, W.; Wang, Y.; Hu, X. A novel AuNPs-doped COFs composite as electrochemical probe for chlorogenic acid detection with enhanced sensitivity and stability. *Sens. Actuators B Chem.* **2018**, *276*, 362–369. [CrossRef]
32. Klingler, R.J.; Kochi, J.K. Electron-transfer kinetics from cyclic voltammetry. Quantitative description of electrochemical reversibility. *J. Phys. Chem.* **1981**, *85*, 1731–1741. [CrossRef]
33. Wang, K.; Liu, Q.; Guan, Q.M.; Wu, J.; Li, H.N.; Yan, J.J. Enhanced direct electrochemistry of glucose oxidase and biosensing for glucose via synergy effect of graphene and CdS nanocrystals. *Biosens. Bioelectron.* **2011**, *26*, 2252–2257. [CrossRef] [PubMed]
34. Moru, S.; Kumar, V.S.; Kummari, S.; Yugender Goud, K. A disposable screen printed electrodes with hexagonal $Ni(OH)_2$ nanoplates embedded chitosan layer for the detection of depression biomarker. *Micromachines* **2023**, *14*, 146. [CrossRef] [PubMed]

35. Lin, K.C.; Hong, C.P.; Chen, S.M. Simultaneous determination for toxic ractopamine and salbutamol in pork sample using hybrid carbon nanotubes. *Sens. Actuators B Chem.* **2013**, *177*, 428–436. [CrossRef]
36. Kong, L.J.; Pan, M.F.; Fang, G.Z.; He, X.L.; Yang, Y.K.; Dai, J.; Wang, S. Molecularly imprinted quartz crystal microbalance sensor based on poly(o-aminothiophenol) membrane and Au nanoparticles for ractopamine determination. *Biosens. Bioelectron.* **2014**, *51*, 286–292. [CrossRef]
37. Li, J.J.; Bo, X.J. Laser-enabled flexible electrochemical sensor on finger for fast food security detection. *J. Hazard. Mater.* **2022**, *423*, 127014. [CrossRef]
38. Luo, S.X.; Wu, Y.H.; Gou, H.; Liu, Y. A novel electrochemical sensor for the analysis of salbutamol in pork samples by using $NiFe_2O_4$ nanoparticles modified glassy carbon electrode. *Adv. Mater. Res.* **2013**, *850–851*, 1279–1282. [CrossRef]
39. Wang, Q.W.; Wang, M.; Zhang, N.; Huang, X.; Wang, X.; Wang, S. A novel electrochemical sensor based on MoS_2 electrospun nanofibers and polyoxometalate composite for the simultaneous detection of ractopamine and clenbuterol. *Microchem. J.* **2023**, 108434. [CrossRef]

foods

MDPI

Article

Cultivation Factors That Affect Amyloid-β Aggregation Inhibitory Activity in *Perilla frutescens* var. *crispa*

Keiya Shimamori [1], Tomohiko Nambu [1], Daiki Kawamata [1], Masahiro Kuragano [1], Naoki Nishishita [2], Toshifumi Iimori [1], Shinya Yamanaka [1], Koji Uwai [1] and Kiyotaka Tokuraku [1,*]

[1] Graduate School of Engineering, Muroran Institute of Technology, Muroran 050-8585, Japan
[2] Regenerative Medicine and Cell Therapy Laboratories, Kaneka Corporation, Kobe 650-0047, Japan
* Correspondence: tokuraku@mmm.muroran-it.ac.jp; Tel.: +81-0143-46-5721

Abstract: Alzheimer's disease (AD) is thought to be caused by the deposition of amyloid-β (Aβ) in the brain. Aβ begins to aggregate approximately 20 years before the expression of its symptoms. Previously, we developed a microliter-scale high-throughput screening (MSHTS) system for inhibitors against Aβ aggregation using quantum dot nanoprobes. Using this system, we also found that plants in the *Lamiaceae*, particularly *Perilla frutescens* var. *crispa*, have high activity. The cultivation environment has the potential to enhance Aβ aggregation inhibitory activity in plants by changing their metabolism. Here, we report on cultivation factors that affected the activity of *P. frutescens* var. *crispa* cultivated in three fields under different cultivation conditions. The results revealed that the activity of *P. frutescens* var. *crispa* harvested just before flowering was highest. Interestingly, the activity of wind-shielded plants that were cultivated to prevent exposure to wind, was reduced to 1/5th of plants just before flowering. Furthermore, activity just before flowering increased following appropriate nitrogen fertilization and at least one week of drying from the day before harvest. In addition, we confirmed that the *P. frutescens* var. *crispa* leaf extracts suppressed Aβ-induced toxicity in nerve growth factor-differentiated PC12 cells. In this study, we demonstrated that flowering, wind, soil water content, and soil nitrogen content affected Aβ aggregation inhibitory activity, necessary to suppress Aβ neurotoxicity, in *P. frutescens* var. *crispa* extracts. This study provides practical cultivation methods for *P. frutescens* var. *crispa* with high Aβ aggregation inhibitory activity for the prevention of AD.

Keywords: Alzheimer's disease; amyloid-β; quantum dot; *Perilla frutescens* var. *crispa*; cultivation method; fertilization

1. Introduction

Alzheimer's disease (AD) is the most common form of dementia and an acute medical issue due to the increasing number of AD patients worldwide. It is estimated that the number of people with dementia will increase from 57.4 million worldwide in 2019 to 152.8 million cases in 2050 [1]. Furthermore, the value of a statistical life-based global economic burden of Alzheimer's disease and related dementias was estimated at USD 2.8 trillion in 2019 and was projected to increase to USD 16.9 trillion in 2050 [2]. AD progressively causes memory loss and cognitive dysfunction, eventually requiring full-time care [3]. Most studies target amyloid-β (Aβ) peptides that mainly consist of 40 or 42 amino residues [4]. Aβ grows into oligomers, fibrils, and aggregates and is observed as plaques in a patient's brain [5]. Thus, Aβ aggregation is an initiating factor in AD that leads to the death of neurons [6,7]. Therefore, many studies have been performed aimed at decreasing Aβ aggregation [8].

Nevertheless, most clinical trials targeting AD patients failed [9]. These results might stem from Aβ aggregation that begins about two decades before the onset of symptoms [10], suggesting that treatment for patients in whom AD has developed is too late. Recently, a few clinical trials achieved better results using a treatment based on anti-Aβ antibodies in patients with mild cognitive impairment (MCI) and mild AD. Primarily, the US Food and Drug Administration approved aducanumab, which is a therapeutic medicine based on anti-Aβ fibril antibodies, via an accelerated approval pathway [11]. The treatment of patients with MCI and mild AD with aducanumab showed the removal of Aβ in the brain and reduced AD pathology [12]. In addition, treatment with lecanemab, an anti-Aβ protofibril antibody, showed an even greater reduction in cognitive decline in patients with MCI and mild AD [13]. These achievements support the concept that it is important to inhibit Aβ aggregation before the onset of AD symptoms. However, treatments with antibodies are very expensive [14] and are thus not accessible to most people.

Even if Aβ aggregates are removed after the neural network is severely damaged, it is expected that the network will not recover to its initial state due to its irreversibility. We believe that it is important to prevent neural networks from being damaged. In addition to treatment by removing Aβ aggregation with anti-Aβ antibody drugs, we suggest the importance of adding a preventive effect against Aβ aggregation to the daily diet. Since food enables us to take any substance affordably and permanently, we assume that food with Aβ aggregation inhibitory activity contributes greatly to AD prevention. Previously, we developed a microliter-scale high-throughput screening (MSHTS) system to evaluate $A\beta_{42}$ aggregation inhibitors in vitro by using quantum dot (QD) nanoprobes [15]. The MSHTS system enables high-throughput screening and the accurate evaluation of samples, such as foods containing various contaminants [16–19]. Generally, atomic force microscopy and electron microscopy (EM) have been used to directly observe Aβ fibrils [20]. In particular, the latest cryo-EM provides high-resolution images of the structure of amyloid fibrils [21]. Although these methods are powerful tools for observing the structure of fibrils, they are not suited for quantification and high-throughput screening. Generally, Aβ aggregation is quantified using thioflavin T (ThT) [22], which measures the amount of fluorescence bound to the β-sheet. However, the ThT assay cannot accurately evaluate activity in contaminated samples because fluorescence is inhibited by many dyes in the samples [16]. In contrast, the MSHTS system allows the accurate evaluation of total Aβ inhibitory activity in various samples, such as crude extracts.

Using the MSHTS system, we have already evaluated the Aβ aggregation inhibitory activity of commercial dressings, which are processed foods, and reported those findings in this journal [16]. Furthermore, we applied this system to food plants and found that one family in particular, the *Lamiaceae*, displayed high activity [15]. In addition, we attempted to isolate active compounds from summer savory, a member of the *Lamiaceae*, revealing that the main active compound was rosmarinic acid (RA), a polyphenol. Subsequently, we found that Aβ aggregation inhibitory activity in leaf extracts from *Perilla frutescens* var. *crispa* was higher than from spearmint (*Mentha spicata*), and had the highest activity among all tested members of the *Lamiaceae* [15]. Presumably, this high activity derives from its rich polyphenol content [23–25]. However, we preliminarily confirmed that the activity differed, depending on harvest time and cultivation practice, even within *P. frutescens* var. *crispa*. Even in the same plants, the content of polyphenols changed due to the growth cycle and cultivation conditions, such as fertilization [26–28], including in perilla [29–31].

In this study, we investigated the cultivation factors that affected the Aβ aggregation inhibitory activity of *P. frutescens* var. *crispa*. We evaluate total activity in extracts using the automated MSHTS system, which allows for high-throughput screening [18]. While some papers have indicated cultivation methods for *P. frutescens* var. *crispa* to increase the content of polyphenols and other promising bioactive compounds [28,30], it is unclear whether those contents would contribute to total Aβ aggregation inhibitory activity. Here, we evaluated the activity of *P. frutescens* var. *crispa* cultivated in different fields and fertilization conditions. Consequently, we found that flowering, wind, nitrogen fertilization, and water content affected the activity of *P. frutescens* var. *crispa*. This study provides practical cultivation methods for enhancing Aβ aggregation inhibitory activity.

2. Materials and Methods

2.1. Materials

Human $A\beta_{42}$ (4349-v, Peptide Institute Inc., Osaka, Japan) and Cys-conjugated $A\beta_{40}$ (23519, Anaspec Inc., Fremont, CA, USA) were purchased commercially. Thiazolyl blue tetrazolium bromide (MTT)(M5655) and poly-D-lysine were purchased from Sigma-Aldrich (St. Louis, MO, USA). Nerve growth factor (NGF) was purchased from Cosmo Bio (Toyo, Japan).

2.2. Preparation of QDAβ Nanoprobe

The QDAβ nanoprobe was prepared using QD-PEG-NH_2 (Qdot^{TM} 605 ITK^{TM} Amino (PEG) Quantum dot; Q21501MP, Waltham, MA, USA, Thermo Fisher Scientific) according to our previous reports [15,18,32]. QDAβ concentration was determined by comparing absorbance at 350 nm to unlabeled QD-PEG-NH_2.

2.3. Evaluation Using the Automated MSHTS System

The half maximal effective concentration (EC_{50}) values of all plant extracts were determined by a modified automated MSHTS system, as was described in our previous report [18]. Specifically, mixed solutions were prepared with six concentrations of extracts, 25 nM QDAβ, and 25 μM $A\beta_{42}$ in PBS containing 5% ethanol (EtOH) and 2.5% dimethyl sulfoxide (DMSO) and incubated in a 1536-well plate (782096, Greiner, Kremsmünster, Austria) at 37 °C for 24 h. The images of each well were captured pre- and post-incubation by an inverted fluorescence microscope (see Section 2.4.). Standard deviation (SD) values of the central images of the region of interest (432 × 432 pixels) in each well were measured by the General Analysis program of NIS-Elements (Nikon, Tokyo, Japan). The images on the highest extract concentration, including insoluble substances, were eliminated because these affect SD values and disrupt accurate evaluation. EC_{50} was estimated from the SD values by Prism software (GraphPad software, San Diego, CA, USA) using an EC_{50} shift by global fitting (asymmetric sigmoidal, five-parameter logistic) [15].

2.4. Fluorescence Microscopy

Fluorescence images in the 1536-well plate were observed by an inverted fluorescence microscope system (ECLIPSE Ti-E, Nikon) equipped with a color CMOS camera (DS-Ri2, Nikon). QD fluorescence was imaged using a 4× objective lens (MRD00045, Nikon) and a TRITC filter set (TRITC- A-Basic-NTE, Semrock, NY, USA).

2.5. MTT Assay

Rat adrenal pheochromocytoma, PC12 cells, were obtained from the JCRB Cell Bank (Osaka, Japan). PC12 cells were maintained in dulbecco's modified eagle medium supplemented with 10% fetal bovine serum (FBS), 100 U/mL penicillin, and 100 μg/mL streptomycin, as was described in our previous report [33]. Cells were cultured at 37 °C in humidified air containing 5% CO_2 with CO_2 incubator. PC12 cells were plated at 1.2×10^4 cells in poly-D-lysine-coated 96-well plates (3860-096, AGC TECHNO GLASS, Shizuoka, Japan) and incubated for 24 h. After incubation, cells were treated with 50 ng/mL NGF for 24 h for neuronal differentiation. After this step, cells were cultured in a medium without FBS but containing 50 ng/mL NGF. Cells were treated with extracts and 25 μM $A\beta_{42}$ for 24 h by replacing the medium with 1% EtOH/1% DMSO. Cells were then treated with 0.5 mM MTT for 4 h by medium replacement in reference to the MTT assay [34]. After incubation, the supernatant of each well was removed, and formazan crystals were dissolved in a 10% SDS/0.01 M HCl solution. After overnight incubation, absorbance (570 nm) was measured in each well with (SH-9000Lab, CORONA ELECTRIC, Ibaraki, Japan). Cell viability was calculated as a percentage relative to cells treated with the control medium (1% EtOH/1% DMSO).

2.6. Cultivation of *Perilla frutescens* var. *crispa*

P. frutescens var. *crispa* seedlings were purchased from AW Farm Chitose Co., Ltd. (Hokkaido, Japan). Seeds were planted around May and harvested regularly. In 2018, plants were cultivated in Chitose (42°53′24.5″ N 141°41′41.5″ E), Watenbetsu (42°58′08.8″ N 144°00′02.9″ E), and Tomaribetsu (43°01′48.1″ N 144°07′55.2″ E) in Hokkaido, and in 2019, on the field on a college campus (42°22′38.1″ N 141°02′12.1″ E). In 2020, plants were cultivated in a plastic greenhouse on a college campus. In 2018, meteorological data was measured with meteorological equipment installed in each field. In 2019 and 2020, soil from the plot was sampled at the same time as the harvest from the field on the college campus. The water content of the soil was calculated from pre- and post-drying weights.

2.7. Preparation of Extracts

Extracts were prepared by extraction with 5 mL of 95% EtOH per gram of fresh leaf weight at room temperature for seven days in 2018. In 2019/2020, perilla leaf extracts were prepared by extraction for one day to prepare many samples efficiently. After the EtOH solution was filtered and concentrated in vacuo, samples were prepared as a 10 mg/mL extract with EtOH.

2.8. Statistical Analysis

All statistical analyses were performed in Excel (version 16; Microsoft, Redmond, WA, USA). A two-tailed Welch's *t*-test was performed for comparisons between groups. Soil water content and EC_{50} were evaluated by linear regression using least squares in Excel. $p < 0.05$ was considered statistically significant (*: $p < 0.05$, **: $p < 0.01$, ***: $p < 0.005$).

3. Results

3.1. Principle of the MSHTS System

We evaluated more than a hundred samples of *P. frutescens* var. *crispa* extracts for Aβ aggregation inhibitory activity using the automated MSHTS System [18] (representative; Figure 1B,C). This system's uniqueness is in its use of QDs, which are nanoscale crystals with an overwhelmingly non-photobleaching property (Figure 1A). After incubation of a mixture of the fluorescent probe QDAβ and $A\beta_{42}$, QDAβ is incorporated into $A\beta_{42}$ fibrils. From the fluorescence images of the mixtures, the variation in brightness was measured (Figure 1B) as an SD value, which can be considered as the progress of aggregation. Aβ aggregation inhibitory activity was evaluated by developing an inhibition curve using SD values (Figure 1C) and calculating the EC_{50}. Lower EC_{50} values indicate higher activity. If the inhibition curve did not reach 50% of the percentage of SD values, then a sample was determined to be not detectable (ND).

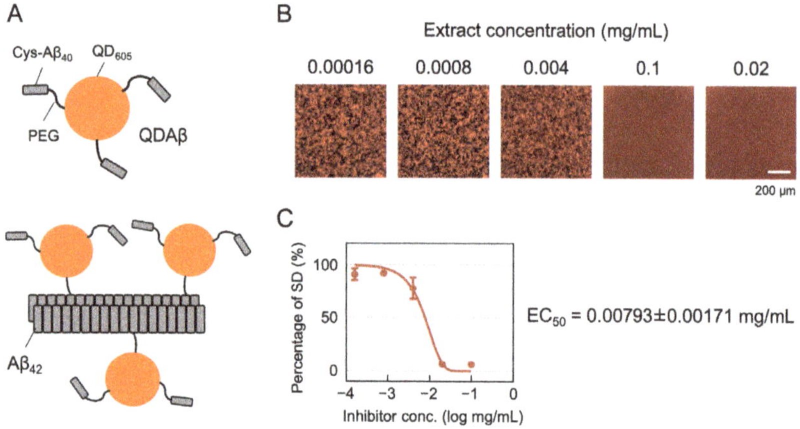

Figure 1. Principle of the MSHTS system for evaluating Aβ_{42} aggregation inhibitory activity. (**A**) QDAβ is a fluorescent probe prepared by cross-linking Cys-Aβ_{40} and QD_{605} via amino polyethylene glycol (PEG)(top). QDAβ co-aggregates with unlabeled Aβ_{42}, allowing Aβ_{42} fibrils to be visualized by fluorescence microscopy (bottom). (**B**,**C**) Representative analysis of Aβ aggregation inhibitory activity of *Perilla frutescens* var. *crispa* leaf extract. Mixed solutions of 25 µM Aβ_{42} and extract were added to 1536-well plates and imaged by fluorescence microscopy before and after incubation at 37 °C for 24 h. (**B**) The fluorescence images were trimmed to 432 x 432 pixels, and the SD values, meaning the variation in brightness, were measured. (**C**) The inhibition curve was drawn from SD values, and EC_{50} was calculated using Prism GraphPad software.

3.2. Aβ Aggregation Inhibitory Activity of Perilla frutescens *var.* crispa *and Meteorological Data in Three Fields*

At first, we cultivated *P. frutescens* var. *crispa* plants in three fields in Hokkaido, Japan, in 2018 to investigate regional differences in Aβ aggregation inhibitory activity and installed meteorological equipment in each field in late August (Figure 2). Plants were harvested on three dates (24 and 25 July, 28 and 29 August, and 5 and 10 October). Extracts from leaves, stems, and roots were compared in late July. Leaf extracts showed high activity in all fields, while stem extracts in Tomaribetsu and root extracts in Chitose and Watenbetsu showed no activity (i.e., ND) (Figure 3A). In Chitose and Watenbetsu, roots showed no activity, and leaves showed significantly higher activity than stems. Therefore, we compared the activity of extracts of leaves harvested on the three dates (Figure 3B); the extracts from plants growing in Chitose and Tomaribetsu fields showed higher activity on October 5 and 10 than on July 24 and 25. Furthermore, we compared the activity of leaves harvested on October 5 and 10 in plants growing in different fields: the highest activity was observed in extracts from plants grown in the Chitose field, followed by the Tomaribetsu and Watenbetsu fields.

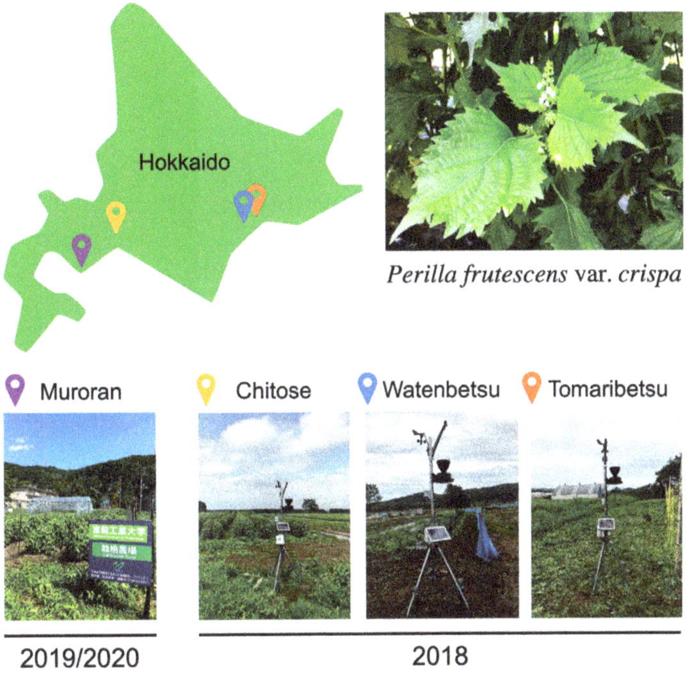

Figure 2. Cultivation of *Perilla frutescens* var. *crispa* in four fields in Hokkaido. In 2018, plants were cultivated in three fields (Chitose, Watenbetsu, and Tomaribetsu) in Hokkaido, Japan, where meteorological equipment was installed. In 2019 and 2020, plants were cultivated under multiple conditions in only Muroran.

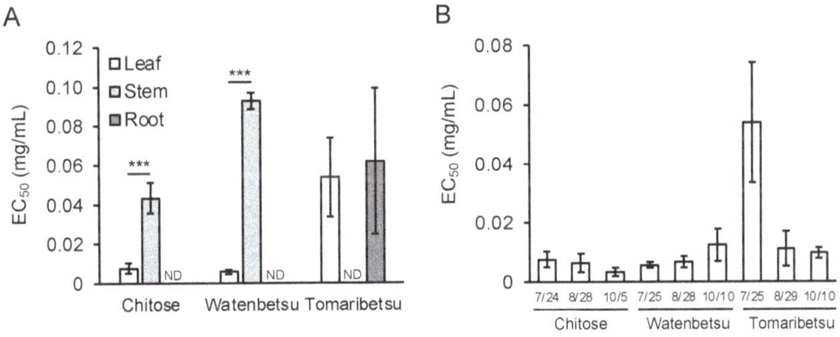

Figure 3. Aβ aggregation inhibitory activity of *Perilla frutescens* var. *crispa* leaves differed depending on fields and harvest time. (**A**) EC_{50} in leaf, stem, and root extracts. Plants were cultivated in three fields in 2018 and harvested in late July. (**B**) The activity of *Perilla frutescens* var. *crispa* leaves was harvested in late July, late August, and early October. The activity was determined to be not detectable (ND) in samples if the inhibitory curve did not reach 50% of the percentage of SD values. Data are represented as mean $\pm$ SD (n = 4 separate experiments). Comparison of EC_{50} values (***: $p <$ 0.005, Welch's t-test).

We also investigated the correlation between meteorological data and Aβ aggregation inhibitory activity. Meteorological data from August 24 to October 4 were acquired from each field and were plotted as a daily average (Figure 4). These data appear to show that low soil water content (Figure 4G) was correlated with the above-mentioned plot-related rank in activity (Chitose < Tomaribetsu < Watenbetsu). Wind speed was highest at Chitose (Figure 4E,F), similar to the activity trend. Thus, we speculated that wind and water content affect Aβ aggregation inhibitory activity.

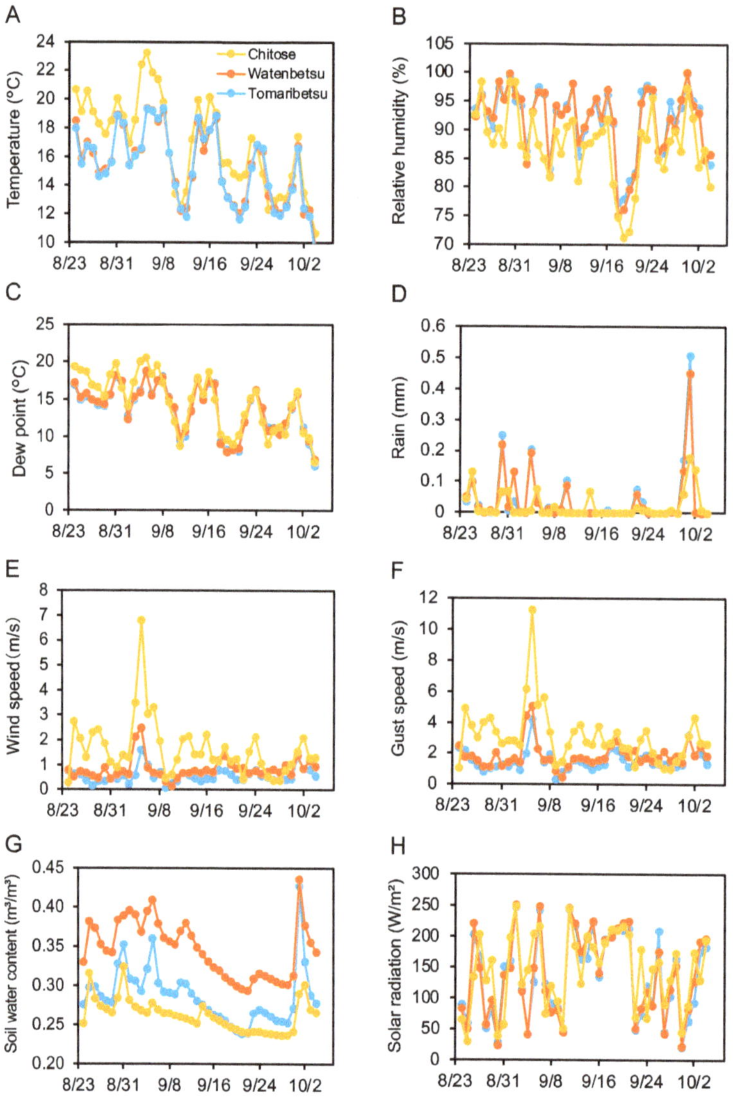

Figure 4. Wind speed and soil water content correlate with regional differences in Aβ aggregation inhibition activity. Meteorological data was measured in three fields from August 24 to October 4, shown as daily averages. Yellow, red, and blue show Chitose, Watenbetsu, and Tomaribetsu, respectively. Temperature (**A**), relative humidity (**B**), dew point (**C**), rain (**D**), wind speed (**E**), gust speed (**F**), soil water content (**G**), and solar radiation (**H**). Data represent an average of each day from every 10 min measurement.

3.3. *Aβ Aggregation Inhibitory Activity of Perilla frutescens* var. *crispa Was Highest Just before Flowering and Was Enhanced by the Wind*

In 2019/2020, we prepared extracts efficiently with a change in extraction time from seven days to one day and evaluated many samples to investigate the details of cultivation conditions. We confirmed no change of Aβ aggregation inhibitory activity by extraction time (Supplemental Figure S1). In 2019, plots of various conditions with additional fertilization were made after laying compost on the field in Muroran, and *P. frutescens* var. *crispa* was cultivated (Figure 2). Plants cultivated without additional fertilizer were referred to as the standard (Std) and harvested every four weeks. Activity in the leaves of Std plants was highest just before the flowering, on 12 September, decreasing thereafter (Figure 5A). Plants were cultivated in the same soil conditions as Std, but a windbreak was created, enclosing plants with a plastic sheet (Standard with windbreak; Sw)(Figure 5B). In Sw plants, activity just before flowering was significantly lower than that of Std plants, with an EC_{50} that was five-fold higher (Figure 5A). The activity of Sw plants harvested on October 3 was highest, but EC_{50} was still three-fold higher than Std plants harvested on September 5. These results indicate that wind exposure enhanced Aβ aggregation inhibition activity in *P. frutescens* var. *crispa*.

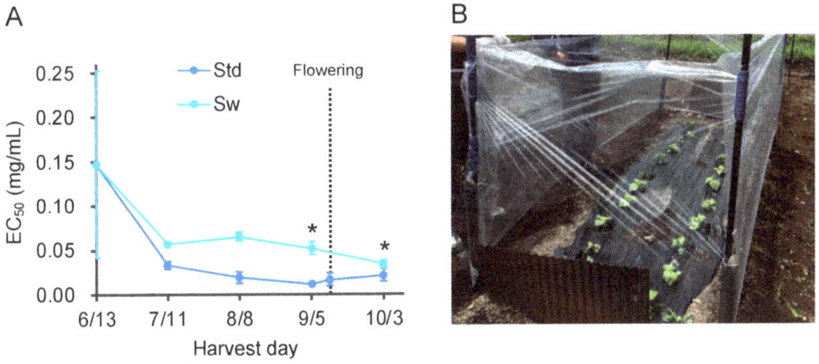

Figure 5. Aβ aggregation inhibitory activity showed the highest just before flowering, with wind enhancement. (**A**) Aβ aggregation inhibitory activity of leaves in regularly harvested *Perilla frutescens* var. *crispa* in 2019. *P. frutescens* var. *crispa* was planted on June 13, and whole plants were harvested every 4 weeks. Plants flowered on 12 September. The Std is the condition with no additional fertilizer under all conditions. Plants from Std were also harvested on September 12, during flowering. (**B**) Plants in the standard with windbreak (Sw) were cultivated in the same soil as Std and enclosed with a plastic sheet. Data represent the mean ± SD (n = 3, separate plant). Comparison of the EC_{50} values vs. Std harvested on September 5 (*: $p < 0.05$, Welch's t-test).

3.4. *Adjustment of Soil Nitrogen and Soil Water Content Enhanced Aβ Aggregation Inhibitory Activity of Perilla frutescens* var. *crispa Just before Flowering*

In 2019, *P. frutescens* var. *crispa* plants were cultivated under fertilization conditions other than Std, as shown in Table 1. Specifically, the activities of four major types of fertilizers were compared: slaked lime and calcium carbonate powder, scallop shell powder (Sp) and granular (Sg) with zeolite, each superphosphate of lime (SPL), ammonium sulfate (AMS), potassium sulfate (SOP), and a combination of Sp, SPL, AMS, and SOP (Figure 6A–D). There were no significant differences between any treatments in the activity of leaves of plants harvested just before flowering. On the other hand, the activity of plants cultivated on soil enhanced with SPL, AMS, and SOP fertilizer, just before flowering, was highest in AMS-fertilized conditions (Figure 6E).

Table 1. Multiple Fertilization Conditions in 2019.

Sample Name	Additional Fertilization	Amount (g/m^2)
Std	-	
SL-1	Slaked lime	55.5
SL-2	Slaked lime	555.2
Cp	Calcium carbonate powder	163.5
Sp	Shell powder	170.3
Sg-1	Shell granular (particle size; 0.50–1.00 mm)	340.7
Sg-2	Shell granular (particle size; 1.00–2.00 mm)	340.7
Sg-3	Shell granular (particle size; 3.35–5.60 mm)	340.7
P	Superphosphate of lime (SPL)	114.3
N	Ammonium sulfate (AMS)	95.2
K	Potassium sulfate (SOP)	40.0
PN	Combination of P and N	
PK	Combination of P and K	
NK	Combination of N and K	
PNK	Combination of P, N, and K	
Sp-PNK	Combination of Sp, P, N, and K	

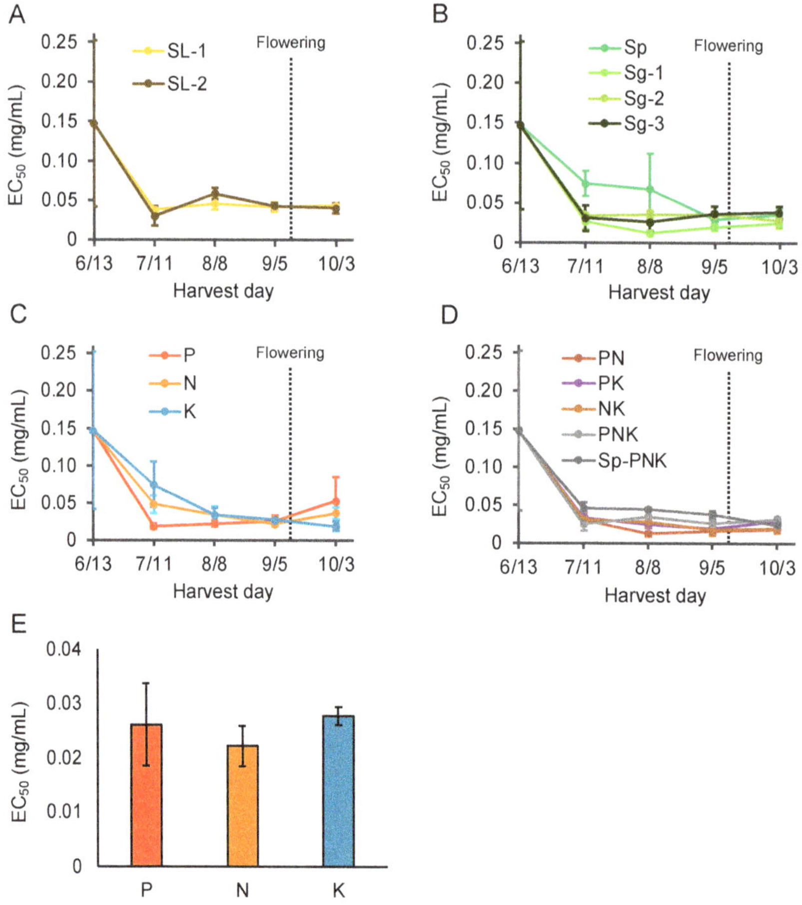

Figure 6. Nitrogen fertilization enhanced the Aβ aggregation inhibitory activity of *Perilla frutescens* var. *crispa* leaves. Plants were cultivated under multiple fertilization conditions and harvested every

4 weeks. (**A–D**) Aβ aggregation inhibitory activity of leaves harvested in 2019. Activity in leaves of plants cultivated in soil with varying amounts of (**A**) slaked lime (SL), (**B**) shell powder (Sp), and shell granular (Sg) fertilizer (Table 1). (**C**,**D**) Activity in leaves of plants cultivated in soil with a combination of ammonium sulfate, lime superphosphate lime, and potassium sulfate fertilizer. (**E**) The activity of (**C**) was compared on September 5, just before flowering. Data represent the mean ± SD (n = 3, separate plants).

In 2020, to identify an appropriate amount of AMS fertilizer, *P. frutescens* var. *crispa* plants cultivated with six levels of AMS fertilization were harvested every four weeks (Table 2). Activity in all six fertilization conditions increased until September 9, just before flowering on September 10 (Figure 7A). Just before flowering, Aβ aggregation inhibitory activity and total fresh leaf weight from each plant were the highest in the N3 treatment (Figure 7B,C). In particular, total leaf weight improved from Std to N3 but decreased above N4. These results show that appropriate nitrogen fertilization increased Aβ aggregation inhibitory activity and total leaf.

Table 2. Fertilizer conditions of nitrogen in 2020.

Sample Name	Amount of Ammonium Sulphate (g/m^2)
Std	0
N1	119
N2	238
N3	476
N4	952
N5	1905

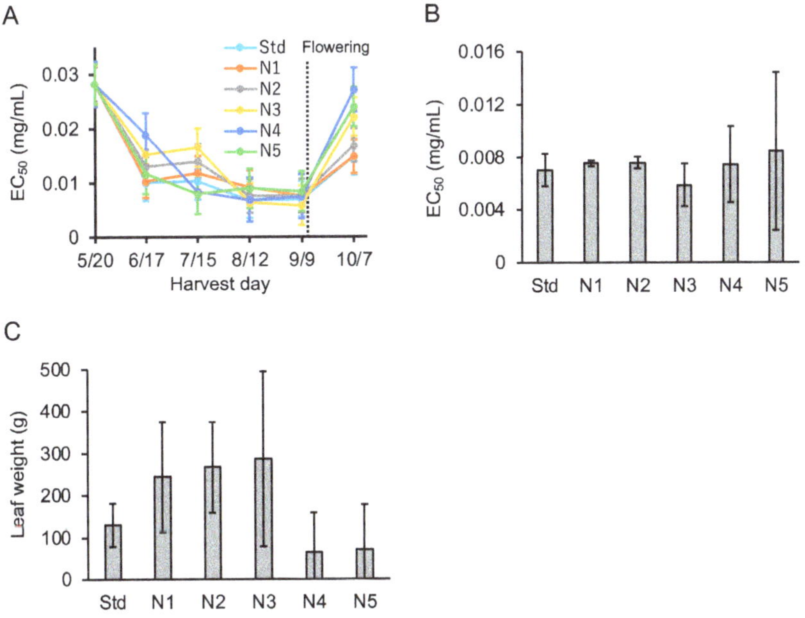

Figure 7. Nitrogen fertilization enhanced Aβ aggregation inhibitory activity and leaf weight. (**A**) Aβ aggregation inhibitory activity in leaves of *Perilla frutescens* var. *crispa* cultivated in a plastic greenhouse with six levels of ammonium sulfate fertilization (Table 2) and harvested every four weeks in 2020. (**B**,**C**) Activity (**B**) and leaf weight (**C**) just before flowering on September 9 in (**A**). Data represent as mean ± SD (n = 3, separate plants).

When *P. frutescens* var. *crispa* plants cultivated in all conditions in 2019 were harvested, soil samples were drawn to analyze soil water content (Table 1). We created a scatter plot between soil water content and the activity of leaves and performed linear regression bordering on 20% water content (Figure 8A). The linear regression shows that activity was high at 20% but decreased markedly when water content exceeded 20%. Therefore, to test whether the cessation of watering just before harvest would enhance activity, plants were cultivated in greenhouses in 2020 and watering was stopped every week starting before the day of harvest. Predictably, soil water content decreased when watering was stopped (Figure 8C). On the other hand, Aβ aggregation inhibition activity was enhanced when watering was stopped for one week and higher than on the day before the harvest date. A scatter plot for this data was created, as shown in Figure 8A, although a linear regression was not performed due to too few samples (Figure 8D). The plots with water content exceeding 20% had highest activity. These results indicate that Aβ aggregation inhibitory activity was enhanced by stopping watering before harvest to reduce soil water content.

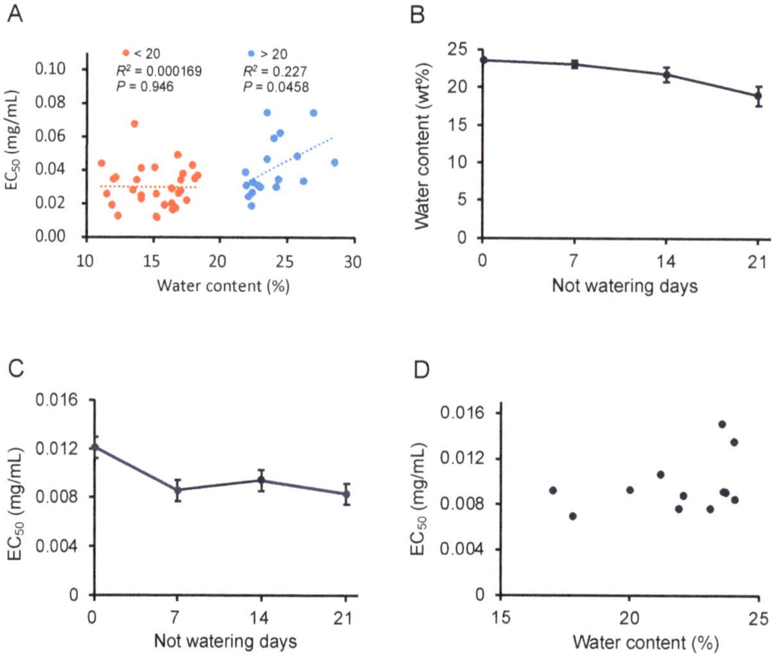

Figure 8. Drying enhanced the Aβ aggregation inhibitory activity of *Perilla frutescens* var. *crispa* leaves. (**A**) Scatter plot of soil water content at harvesting and EC_{50} of leaf extracts from *P. frutescens* var. *crispa* under conditions described in Table 1 harvested on 11 July, 8 August, and 5 September in 2019. Linear regression was performed on more and less than 20% s water content. (**B–D**) Plants were cultivated in greenhouses in 2020, and watering was stopped every week from the day before the harvest date (15 September) to determine their Aβ aggregation inhibitory activity (**C**) and soil water content (**B**). A scatter plot was created in (**D**) about water content and EC_{50} similar to (**A**). (**A**,**D**) represented as mean, and (**B**,**C**) represented as mean ± SD (n = 3, separate plants).

3.5. Suppression Effect of *Perilla frutescens* var. *crispa* on Aβ-Induced Neurocytotoxicity

We tested whether the *P. frutescens* var. *crispa* extract could suppress Aβ-induced neurocytotoxicity by the MTT assay using NGF-differentiated PC12 cells (Figure 9). The data show that 25 μM $A\beta_{42}$ reduced the viability of PC12 cells by 40% relative to the control. The extract significantly restored cell viability and suppressed Aβ-induced neurocytotoxicity the most at 0.781 μg/mL. In contrast, the extracts with more than 25 μg/mL concentrations reduced cell viability to less than 25 μM Aβ and significantly when the concentration was 100 μg/mL.

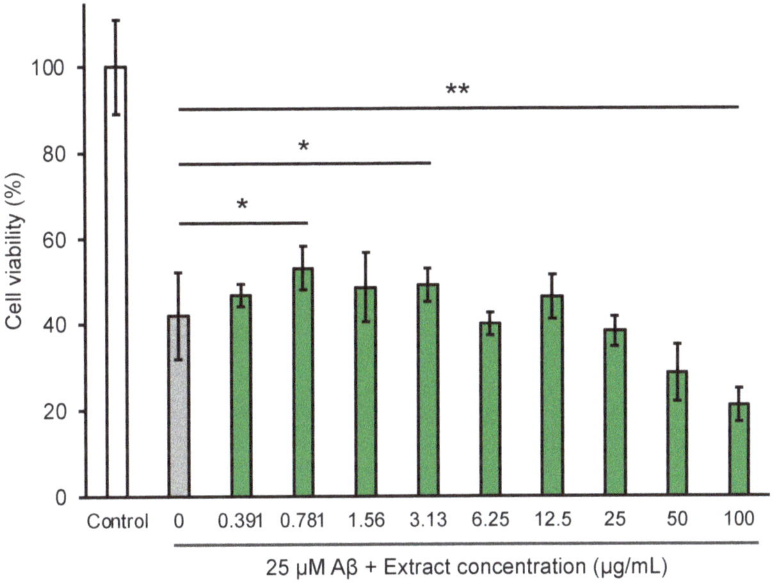

Figure 9. Extracts from *Perilla frutescens* var. *crispa* leaves suppressed Aβ-induced cytotoxicity in NGF-differentiated PC12 cells. PC12 cells were differentiated by NGF for 24 h and treated with 25 μM $A\beta_{42}$ and leaf extract for 24 h. Cell viability was measured by the MTT assay and shown as the relative percentage of absorbance of treated samples compared to the control without Aβ and extract. Each plot and bar graph represent the mean ± SD ($n = 3$ separate experiments with extracts, 6 separate experiments in control, and 25 μM Aβ only). Comparison of the values (*: $p < 0.05$, **: $p < 0.01$, Welch's *t*-test).

4. Discussion

In this study, we evaluated the Aβ aggregation inhibitory activity of *P. frutescens* var. *crispa* cultivated under multiple conditions using the MSHTS system, which has the advantage of directly assessing this activity. At first, we evaluated the activity of plants cultivated in three fields (Figure 3) and found that leaves showed the highest activity among all plant parts that were assessed. In particular, this study's highest activity of N3 (EC_{50} = 0.0059 mg/mL; Figure 7B) was the second highest in the 504 natural plant extracts previously evaluated [18]. We speculate that this activity is due to polyphenols having Aβ inhibitory effect because perilla leaves are rich in polyphenols [23–25]. Several compounds have been reported to show Aβ inhibitory activity among the polyphenols that are found in perilla. In particular, RA is the most promising compound since it is abundant in perilla [23] and has high Aβ aggregation inhibitory activity [15]. Similarly, chalcone [35] and asaron derivatives [36] isolated from perilla have also been reported as having this activity. Since perilla has a wide variety of compounds, total activity is considered by their accumulation and synergistic effect. These secondary metabolites are biosynthesized in response to environmental stress to protect against their oxidation [29,30].

Overall, the Aβ aggregation inhibitory activity of *P. frutescens* var. *crispa* increased until early September, until flowering, and then decreased after flowering (Figures 5A and 7A). In perilla, during flowering, the contents of RA and total phenolic compounds become reduced [28,31]. These contents may have been altered by a change in metabolic pathways.

The amount of total phenolic compounds in plants decreases with flowering due to the oxidation caused by the action of polyphenol oxidase and peroxidase [37]. Therefore, perilla should be harvested just before flowering to avoid a loss of active compounds.

The Aβ aggregation inhibitory activity showed the same trend as wind speed in *Perilla frutescens* var. *crispa* cultivated in the three fields in 2018 (Figures 3 and 4). Furthermore, *Perilla frutescens* var. *crispa* cultivated without wind exposure showed significantly lower activity (Figure 5). Wind, as an environmental stressor, may promote plant growth and affect secondary metabolites. Wind promotes heat exchange, gas trade, and photosynthesis on plant leaves [38]. It was reported that leaf morphology is altered when plants are exposed to wind, increasing the total phenolics in the leaves [39]. Cultivation with more wind may enhance the activity of plants.

In 2019, Aβ aggregation inhibitory activity increased in soil with less than 20% water content in plants cultivated under multiple fertilization conditions (Figure 8A). Furthermore, watering of plants was ceased just before harvest to induce drought stress, and the cessation of watering for more than one week enhanced the activity (Figure 8C). Phenolic acids increase in response to drought stress in plants [40]. Plants biosynthesize polyphenols and flavonoids to defend themselves from drought [41,42]. Abscisic acid (ABA), a plant hormone, is synthesized in response to drought stress in plants [43]. ABA is involved in the accumulation of phenolic acids, especially anthocyanin, in plants under drought stress [44]. Based on the above, we speculate that drought stress promoted the biosynthesis of compounds that contribute to the Aβ aggregation inhibitory activity in *P. frutescens* var. *crispa* via the ABA response pathway.

Aβ aggregation inhibitory activity in *P. frutescens* var. *crispa* was highest with appropriate nitrogen fertilization as N3, and lower activity when excessive fertilization was applied (Figure 7B). In general, plants contain high levels of polyphenols when cultivated under low levels of nitrogen fertilization [45]. It may be essential to adjust fertilization according to nutrient availability in the field. Furthermore, the choice of nitrogen supply reduced the total phenolic content of leaves in strawberries, and the influence is compound-specific [46]. Perilla contains not only the main active compounds, such as RA, but also other active compounds [23]. Nitrogen fertilization might alter the composition of compounds to enhance total activity.

We showed that *P. frutescens* var. *crispa* leaf extracts suppressed Aβ-induced neurocytotoxicity in neuronally differentiated PC12 cells (Figure 9). The methanolic extracts of *Perilla frutescens* (L.) Britton similarly rescued the Aβ-induced cytotoxicity of PC12 cells [47]. Furthermore, compounds isolated from a *Perilla frutescens* (L.) Britton extract also suppressed Aβ-induced cytotoxicity and exhibited toxicity at high concentrations of 100 μg/mL [36]. In this study, the leaf extract at a high concentration of more than 25 μg/mL reduced cell viability but was lower than Aβ toxicity (Figure 9). Thus, an appropriate concentration of perilla extracts is essential so as not to exhibit toxicity.

This study has a limitation: we did not find other compounds that contribute to the Aβ inhibitory activity of *P. frutescens* var. *crispa*. A promising compound seems to be RA because the extracts from fresh perilla leaves prepared similarly in this study contain a high content of RA [48]. Furthermore, we previously reported that RA is the active component of the extract from summer savory (*Satureja hortensis*) in the *Lamiaceae* [15]. Even though we confirmed the presence of RA in the perilla extracts, the content was not enough to assess total activity. Previously, we did not obtain a fraction even equal to total activity, although we attempted to isolate active compounds from *P. frutescens* var. *crispa* extracts. This result shows that the activity in the extracts may be due to the synergistic effect of multiple compounds. Total activity is important when assessing the intake of substances such as food. A known example of this synergistic effect is the co-administration of epigallocatechin and gallic acid to improve learning ability in a mouse model of brain senescence [49]. We speculate that multiple compounds show a total activity in *P. frutescens* var. *crispa* extracts through synergistic effects and are trying to identify those compounds that contribute to this activity. In addition, it is necessary to research whether these compounds have blood–brain barrier permeability. In a rat model, RA reached the brain via the intraperitoneal administration of extracts from *Plectranthus barbatus* (*Lamiaceae*) herbal tea, including RA [50]. Moreover, perilla leaf extracts improved blocked Aβ aggregates-induced memory impairment and reduced Aβ deposits in the hippocampus with systemic administration in 5XFAD mice with AD-linked mutations [51]. In addition to treatment with anti-Aβ antibody drugs (e.g., aducanumab, lecanemab), prevention with foods such as Perilla may protect many people from AD risk. We expect that this study may contribute to preparing the extracts to show improvement against AD-like pathology.

5. Conclusions

In this study, we found several cultivating and harvesting factors of *Perilla frutescens* var. *crispa* that enhanced its Aβ aggregation inhibitory activity. The best harvesting time was just before flowering, and the following cultivation methods were effective: wind, appropriate nitrogen fertilization, and pre-harvest soil drying. These factors may alter the biosynthesis of polyphenols that exhibit this activity, although the active compounds were not identified. Furthermore, we successfully used the MSHTS system to assess total Aβ aggregation inhibitory activity. We consider this method to be a powerful tool for identifying substances with activity in plant-derived food. We hope that the active substances will be discovered and become a preventive and therapeutic agent against AD, for which no cure has yet been found.

Supplementary Materials: The following supporting information can be downloaded at: www.mdpi.com/xxx/s1. Figure S1: Amyloid-β aggregation inhibitory activity is not changed on perilla leaf extracts within one day and seven days of extraction time

Author Contributions: Conceptualization, S.Y., K.U. and K.T.; Methodology, S.Y., K.U. and K.T.; Software, K.S., T.N., D.K. and M.K.; Formal Analysis, K.S., T.N., D.K., N.N. and M.K.; Investigation, K.S., T.N., D.K., M.K. and K.U.; Resources, S.Y., K.U. and K.T.; Data Curation, K.S. and T.N.; Writing—Original Draft Preparation, K.S.; Writing—Review & Editing, M.K., N.N., T.I., S.Y., K.U. and K.T.; Visualization, K.S. and T.N.; Supervision, K.T.; Funding Acquisition, K.S, S.Y., K.U. and K.T. All authors have read and agreed to the published version of the manuscript.

Funding: This study was funded by JSPS KAKENHI Grant Number JP16H03288 (K.T.), JP25350974 (K.T.), and JP16K01909 (K.U.) and supported by JST SPRING Grant Number JPMJSP2153.

Institutional Review Board Statement: Not applicable.

Informed Consent Statement: Not applicable.

Data Availability Statement: Data is available upon request from the corresponding authors.

Acknowledgments: We are grateful to AW Farm Chitose Co., Ltd., Ohmae Giken Industry Co., Ltd., and Shiranuka town office for the cultivation of *Perilla frutescens* var. *crispa* in each field (Chitose, Watenbetsu, and Tomaribetsu, respectively).

Conflicts of Interest: Naoki Nishishita is an employee of KANEKA Corporation. Kaneka Corporation provides joint research funding to Koji Uwai and Kiyotaka Tokuraku for another study. Therefore, the funding source had no role in the design of the experiment and in providing any materials and equipment. Naoki Nishishita supported the analysis and writing of the paper as an expert on this research area.

References

1. GBD 2019 Dementia Forecasting Collaborators. Estimation of the Global Prevalence of Dementia in 2019 and Forecasted Prevalence in 2050: An Analysis for the Global Burden of Disease Study 2019. *Lancet Public Health* **2022**, *7*, e105–e125. [CrossRef] [PubMed]
2. Nandi, A.; Counts, N.; Chen, S.; Seligman, B.; Tortorice, D.; Vigo, D.; Bloom, D.E. Global and Regional Projections of the Economic Burden of Alzheimer's Disease and Related Dementias from 2019 to 2050: A Value of Statistical Life Approach. *eClinicalMedicine* **2022**, *51*, 101580. [CrossRef] [PubMed]

3. Brodaty, H.; Donkin, M. Family Caregivers of People with Dementia. *Dialogues Clin. Neurosci.* **2009**, *11*, 217–228. [CrossRef] [PubMed]
4. Long, J.M.; Holtzman, D.M. Alzheimer Disease: An Update on Pathobiology and Treatment Strategies. *Cell* **2019**, *179*, 312–339. [CrossRef]
5. Masters, C.L.; Simms, G.; Weinman, N.A.; Multhaup, G.; McDonald, B.L.; Beyreuther, K. Amyloid Plaque Core Protein in Alzheimer Disease and Down Syndrome. *Proc. Natl. Acad. Sci. USA* **1985**, *82*, 4245–4249. [CrossRef]
6. Selkoe, D.J.; Hardy, J. The Amyloid Hypothesis of Alzheimer's Disease at 25 Years. *EMBO Mol. Med.* **2016**, *8*, 595–608. [CrossRef]
7. Hampel, H.; Hardy, J.; Blennow, K.; Chen, C.; Perry, G.; Kim, S.H.; Villemagne, V.L.; Aisen, P.; Vendruscolo, M.; Iwatsubo, T.; et al. The Amyloid-β Pathway in Alzheimer's Disease. *Mol. Psychiatry* **2021**, *26*, 5481–5503. [CrossRef]
8. Panza, F.; Lozupone, M.; Logroscino, G.; Imbimbo, B.P. A Critical Appraisal of Amyloid-β-Targeting Therapies for Alzheimer Disease. *Nat. Rev. Neurol.* **2019**, *15*, 73–88. [CrossRef]
9. Yiannopoulou, K.G.; Anastasiou, A.I.; Zachariou, V.; Pelidou, S.-H. Reasons for Failed Trials of Disease-Modifying Treatments for Alzheimer Disease and Their Contribution in Recent Research. *Biomedicines* **2019**, *7*, 97. [CrossRef]
10. Jack, C.R., Jr.; Knopman, D.S.; Jagust, W.J.; Shaw, L.M.; Aisen, P.S.; Weiner, M.W.; Petersen, R.C.; Trojanowski, J.Q. Hypothetical Model of Dynamic Biomarkers of the Alzheimer's Pathological Cascade. *Lancet Neurol.* **2010**, *9*, 119–128. [CrossRef]
11. Dunn, B.; Stein, P.; Cavazzoni, P. Approval of Aducanumab for Alzheimer Disease-The FDA's Perspective. *JAMA Intern. Med.* **2021**, *181*, 1276–1278. [CrossRef] [PubMed]
12. Budd Haeberlein, S.; Aisen, P.S.; Barkhof, F.; Chalkias, S.; Chen, T.; Cohen, S.; Dent, G.; Hansson, O.; Harrison, K.; von Hehn, C.; et al. Two Randomized Phase 3 Studies of Aducanumab in Early Alzheimer's Disease. *J. Prev. Alzheimers Dis.* **2022**, *9*, 197–210. [CrossRef] [PubMed]
13. van Dyck, C.H.; Swanson, C.J.; Aisen, P.; Bateman, R.J.; Chen, C.; Gee, M.; Kanekiyo, M.; Li, D.; Reyderman, L.; Cohen, S.; et al. Lecanemab in Early Alzheimer's Disease. *N. Engl. J. Med.* **2022**, *388*, 9–21. [CrossRef] [PubMed]
14. Fleming, W.K.; Brown, C.R.; Shrank, W.H. Costly New Alzheimer Disease Medications on the Horizon-Financing Alternatives for Medicare. *JAMA Health Forum* **2020**, *1*, e201148. [CrossRef] [PubMed]
15. Ishigaki, Y.; Tanaka, H.; Akama, H.; Ogara, T.; Uwai, K.; Tokuraku, K. A Microliter-Scale High-Throughput Screening System with Quantum-Dot Nanoprobes for Amyloid-β Aggregation Inhibitors. *PLoS ONE* **2013**, *8*, e72992. [CrossRef]
16. Kuragano, M.; Yoshinari, W.; Lin, X.; Shimamori, K.; Uwai, K.; Tokuraku, K. Evaluation of Amyloid B42 Aggregation Inhibitory Activity of Commercial Dressings by A Microliter-Scale High-Throughput Screening System Using Quantum-Dot Nanoprobes. *Foods* **2020**, *9*, 825. [CrossRef]
17. Lin, X.; Watanabe, K.; Kuragano, M.; Kurotaki, Y.; Nakanishi, U.; Tokuraku, K. Dietary Intake of Rosmarinic Acid Increases Serum Inhibitory Activity in Amyloid A Aggregation and Suppresses Deposition in the Organs of Mice. *Int. J. Mol. Sci.* **2020**, *21*, 6031. [CrossRef]
18. Sasaki, R.; Tainaka, R.; Ando, Y.; Hashi, Y.; Deepak, H.V.; Suga, Y.; Murai, Y.; Anetai, M.; Monde, K.; Ohta, K.; et al. An Automated Microliter-Scale High-Throughput Screening System (MSHTS) for Real-Time Monitoring of Protein Aggregation Using Quantum-Dot Nanoprobes. *Sci. Rep.* **2019**, *9*, 2587. [CrossRef]
19. Ogara, T.; Takahashi, T.; Yasui, H.; Uwai, K.; Tokuraku, K. Evaluation of the Effects of Amyloid β Aggregation from Seaweed Extracts by a Microliter-Scale High-Throughput Screening System with a Quantum Dot Nanoprobe. *J. Biosci. Bioeng.* **2015**, *120*, 45–50. [CrossRef]
20. Adamcik, J.; Mezzenga, R. Study of Amyloid Fibrils via Atomic Force Microscopy. *Curr. Opin. Colloid Interface Sci.* **2012**, *17*, 369–376. [CrossRef]
21. Yang, Y.; Arseni, D.; Zhang, W.; Huang, M.; Lövestam, S.; Schweighauser, M.; Kotecha, A.; Murzin, A.G.; Peak-Chew, S.Y.; Macdonald, J.; et al. Cryo-EM Structures of Amyloid-β 42 Filaments from Human Brains. *Science* **2022**, *375*, 167–172. [CrossRef]
22. LeVine, H. [18] Quantification of β-Sheet Amyloid Fibril Structures with Thioflavin T. In *Methods in Enzymology*; Academic Press: Cambridge, MA, USA, 1999; Volume 309, pp. 274–284.
23. Wang, Z.; Tu, Z.; Xie, X.; Cui, H.; Kong, K.W.; Zhang, L. Perilla Frutescens Leaf Extract and Fractions: Polyphenol Composition, Antioxidant, Enzymes (α-Glucosidase, Acetylcholinesterase, and Tyrosinase) Inhibitory, Anticancer, and Antidiabetic Activities. *Foods* **2021**, *10*, 315. [CrossRef]
24. Tipsuwan, W.; Chaiwangyen, W. Preventive Effects of Polyphenol-Rich Perilla Leaves on Oxidative Stress and Haemolysis. *Sci. Asia.* **2018**, *44*, 162–169. [CrossRef]
25. Ghimire, B.K.; Yoo, J.H.; Yu, C.Y.; Kim, S.-H.; Chung, I.-M. Profiling Volatile and Phenolic Compound Composition and Characterization of the Morphological and Biological Activities of Perilla Frutescence Britton Var. Japonica Accessions. *Acta Physiol. Plant.* **2019**, *41*, 108. [CrossRef]
26. Luță, E.A.; Biță, A.; Moroșan, A.; Mihaiescu, D.E.; Ghica, M.; Mihai, D.P.; Olaru, O.T.; Deculescu-Ioniță, T.; Duțu, L.E.; Popescu, M.L.; et al. The Influence of Phytosociological Cultivation and Fertilization on Polyphenolic Content of Menthae and Melissae Folium and Evaluation of Antioxidant Properties through In Vitro and In Silico Methods. *Plants* **2022**, *11*, 2398. [CrossRef] [PubMed]
27. Suwa, R.; Tajima, H.; Gima, S.; Uehara, N.; Watanabe, K.; Yabuta, S.; Tominaga, J.; Kawamitsu, Y. Polyphenol Production in *Peucedanum Japonicum* Thunb. Varies with Soil Type and Growth Stage. *Hortic. J.* **2018**, *87*, 382–388. [CrossRef]

28. Gai, F.; Peiretti, P.G.; Karamać, M.; Amarowicz, R. Changes in the Total Polyphenolic Content and Antioxidant Capacities of Perilla (*Perilla frutescens* L.) Plant Extracts during the Growth Cycle. *J. Food Qual.* **2017**, *2017*, 7214747. [CrossRef]
29. Meng, L.; Lozano, Y.; Bombarda, I.; Gaydou, E.M.; Li, B. Polyphenol Extraction from Eight Perilla Frutescens Cultivars. *C. R. Chim.* **2009**, *12*, 602–611. [CrossRef]
30. Lu, N.; Bernardo, E.L.; Tippayadarapanich, C.; Takagaki, M.; Kagawa, N.; Yamori, W. Growth and Accumulation of Secondary Metabolites in Perilla as Affected by Photosynthetic Photon Flux Density and Electrical Conductivity of the Nutrient Solution. *Front. Plant Sci.* **2017**, *8*, 708. [CrossRef]
31. Gaihre, Y.R.; Tsuge, K.; Hamajima, H.; Nagata, Y.; Yanagita, T. The Contents of Polyphenols in *Perilla Frutescens* (L.) Britton Var. *Frutescens* (Egoma) Leaves Are Determined by Vegetative Stage, Spatial Leaf Position, and Timing of Harvesting during the Day. *J. Oleo Sci.* **2021**, *70*, 855–859.
32. Tokuraku, K.; Marquardt, M.; Ikezu, T. Real-Time Imaging and Quantification of Amyloid-Beta Peptide Aggregates by Novel Quantum-Dot Nanoprobes. *PLoS ONE* **2009**, *4*, e8492. [CrossRef] [PubMed]
33. Kuragano, M.; Yamashita, R.; Chikai, Y.; Kitamura, R.; Tokuraku, K. Three-Dimensional Real Time Imaging of Amyloid β Aggregation on Living Cells. *Sci. Rep.* **2020**, *10*, 9742. [CrossRef] [PubMed]
34. Mairuae, N.; Connor, J.R.; Buranrat, B.; Lee, S.Y. *Oroxylum indicum* (L.) Extract Protects Human Neuroblastoma SH-SY5Y Cells against Β-amyloid-induced Cell Injury. *Mol. Med. Rep.* **2019**, *20*, 1933–1942. [CrossRef] [PubMed]
35. Murakami, K.; Sakaguchi, Y.; Taniwa, K.; Izuo, N.; Hanaki, M.; Kawase, T.; Hirose, K.; Shimizu, T.; Irie, K. Lysine-Targeting Inhibition of Amyloid β Oligomerization by a Green Perilla-Derived Metastable Chalcone in Vitro and in Vivo. *RSC Chem. Biol.* **2022**, *3*, 1380–1396. [CrossRef]
36. Lee, J.E.; Kim, N.; Yeo, J.Y.; Seo, D.-G.; Kim, S.; Lee, J.-S.; Hwang, K.W.; Park, S.-Y. Anti-Amyloidogenic Effects of Asarone Derivatives From Perilla Frutescens Leaves against Beta-Amyloid Aggregation and Nitric Oxide Production. *Molecules* **2019**, *24*, 4297. [CrossRef]
37. André, A.; Leupin, M.; Kneubühl, M.; Pedan, V.; Chetschik, I. Evolution of the Polyphenol and Terpene Content, Antioxidant Activity and Plant Morphology of Eight Different Fiber-Type Cultivars of *Cannabis sativa* L. Cultivated at Three Sowing Densities. *Plants* **2020**, *9*, 1740. [CrossRef]
38. Mezour, S. Forests Phenolic Compounds Instigate Cloud Precipitation. *Res. Sq.* 2020; preprint. [CrossRef]
39. McArthur, C.; Bradshaw, O.S.; Jordan, G.J.; Clissold, F.J.; Pile, A.J. Wind Affects Morphology, Function, and Chemistry of Eucalypt Tree Seedlings. *Int. J. Plant Sci.* **2010**, *171*, 73–80. [CrossRef]
40. Selmar, D.; Kleinwächter, M. Stress Enhances the Synthesis of Secondary Plant Products: The Impact of Stress-Related over-Reduction on the Accumulation of Natural Products. *Plant Cell Physiol.* **2013**, *54*, 817–826. [CrossRef]
41. Nakabayashi, R.; Yonekura-Sakakibara, K.; Urano, K.; Suzuki, M.; Yamada, Y.; Nishizawa, T.; Matsuda, F.; Kojima, M.; Sakakibara, H.; Shinozaki, K.; et al. Enhancement of Oxidative and Drought Tolerance in Arabidopsis by Overaccumulation of Antioxidant Flavonoids. *Plant J.* **2014**, *77*, 367–379. [CrossRef]
42. González-Villagra, J.; Kurepin, L.V.; Reyes-Díaz, M.M. Evaluating the Involvement and Interaction of Abscisic Acid and MiRNA156 in the Induction of Anthocyanin Biosynthesis in Drought-Stressed Plants. *Planta* **2017**, *246*, 299–312. [CrossRef] [PubMed]
43. Sauter, A.; Davies, W.J.; Hartung, W. The Long-Distance Abscisic Acid Signal in the Droughted Plant: The Fate of the Hormone on Its Way from Root to Shoot. *J. Exp. Bot.* **2001**, *52*, 1991–1997. [CrossRef] [PubMed]
44. González-Villagra, J.; Cohen, J.D.; Reyes-Díaz, M.M. Abscisic Acid Is Involved in Phenolic Compounds Biosynthesis, Mainly Anthocyanins, in Leaves OfAristotelia Chilensisplants (Mol.) Subjected to Drought Stress. *Physiol. Plant.* **2019**, *165*, 855–866. [CrossRef]
45. Heimler, D.; Romani, A.; Ieri, F. Plant Polyphenol Content, Soil Fertilization and Agricultural Management: A Review. *Eur. Food Res. Technol.* **2017**, *243*, 1107–1115. [CrossRef]
46. Narvekar, A.S.; Tharayil, N. Nitrogen Fertilization Influences the Quantity, Composition, and Tissue Association of Foliar Phenolics in Strawberries. *Front. Plant Sci.* **2021**, *12*, 613839. [CrossRef]
47. Kim, D.-J.; Kim, M.-S.; Kim, S.; Hwang, K.-W.; Park, S.-Y. Anti-Amyloidogenic Effects of Perilla Frutescens Var. Acuta on Beta-Amyloid Aggregation and Disaggregation. *J. Food Biochem.* **2017**, *41*, e12393. [CrossRef]
48. Kagawa, N.; Iguchi, H.; Henzan, M.; Hanaoka, M. Drying the Leaves of Perilla Frutescens Increases Their Content of Anticancer Nutraceuticals. *Food Sci. Nutr.* **2019**, *7*, 1494–1501. [CrossRef] [PubMed]
49. Pervin, M.; Unno, K.; Nakagawa, A.; Takahashi, Y.; Iguchi, K.; Yamamoto, H.; Hoshino, M.; Hara, A.; Takagaki, A.; Nanjo, F.; et al. Blood Brain Barrier Permeability of (-)-Epigallocatechin Gallate, Its Proliferation-Enhancing Activity of Human Neuroblastoma SH-SY5Y Cells, and Its Preventive Effect on Age-Related Cognitive Dysfunction in Mice. *Biochem. Biophys. Rep.* **2017**, *9*, 180–186. [CrossRef] [PubMed]

50. Falé, P.L.V.; Madeira, P.J.A.; Florêncio, M.H.; Ascensão, L.; Serralheiro, M.L.M. Function of *Plectranthus barbatus* Herbal Tea as Neuronal Acetylcholinesterase Inhibitor. *Food Funct.* **2011**, *2*, 130–136. [CrossRef]
51. Cho, E.; Lee, J.; Sin, J.S.; Kim, S.-K.; Kim, C.J.; Park, M.H.; Cho, W.-S.; Moon, M.; Kim, D.H.; Jung, J.W. Effects of Perilla Frutescens Var. Acuta in Amyloid β Toxicity and Alzheimer's Disease-like Pathology in 5XFAD Mice. *Food Chem. Toxicol.* **2022**, *161*, 112847. [CrossRef]

foods

MDPI

Article

Development of a Modified QuEChERS Method Based on Magnetic Multi-Walled Carbon Nanotubes as a Clean-Up Adsorbent for the Analysis of Heterocyclic Aromatic Amines in Braised Sauce Beef

Min Li [1,2], Pengxiang Wang [1], Xu Zhang [1], Hongyu Wang [1], Ke Li [1,2] and Yanhong Bai [1,2,*]

1 College of Food and Bioengineering, Zhengzhou University of Light Industry, Zhengzhou 450001, China
2 Henan Key Laboratory of Cold Chain Food Quality and Safety Control, Zhengzhou 450001, China
* Correspondence: baiyanhong212@163.com

Abstract: Heterocyclic aromatic amines (HAAs) generated during the cooking of meats cause adverse effects on human health. The purpose of the current research was to develop a modified QuEChERS (Quick, Easy, Cheap, Effective, Rugged, Safe) method using magnetic multi-walled carbon nanotubes (Fe_3O_4-MWCNTs) as clean-up adsorbents for the rapid determination of HAAs in braised sauce beef. The significant parameters in extraction and clean-up processes were screened and optimized. Under optimal conditions, the LODs ranged from 3.0 ng/g to 4.2 ng/g. The recoveries (78.5–103.2%) and relative standard deviations RSDs (<4.6%) of five HAAs were obtained. These are in accordance with the validation criteria (recovery in the range of 70–120% with RSD less than 20%). Compared with conventional clean-up adsorbents (PSA or C18), Fe_3O_4-MWCNTs displayed equivalent or better matrix removal efficiency, while making the pretreatment process easier and more time-saving through magnetic separation. Less usage of adsorbent makes the method possess another advantage of being lower in cost per sample. The method developed was successfully applied to analyze real samples collected from local deli counters, demonstrating Fe_3O_4-MWCNTs could be considered as an effective alternative adsorbent with great potential in the QuEChERS process.

Keywords: heterocyclic aromatic amines; QuEChERS; Fe_3O_4-MWCNTs; braised sauce beef

1. Introduction

Heterocyclic aromatic amines (HAAs), one family of chemical substances with a heterocyclic structure, are mainly produced from high-protein foods (seafood and meat products) which are cooked at high temperatures. Currently, over 30 HAAs have been found and structurally identified in different cooked foodstuffs [1,2]; several of them are considered to be highly potential carcinogens and mutagens, to which humans are primarily exposed through diet [3,4]. Thus, the food safety issues of HAAs have received much attention by both consumers and researchers.

Currently, various analytical methods, including gas chromatography–mass spectrometry (GC–MS), liquid chromatography–mass spectrometry (LC–MS), liquid chromatography with fluorescence detection (LC–FLD) and/or diode array detection (LC–DAD), have been commonly employed to analyze HAAs [5–7]. Due to the trace amounts of HAAs and the complex food matrix, an appropriate sample pretreatment procedure before analysis is required. Currently, the most frequently used pretreatment methods for HAA analyses are solid-phase extraction (SPE) and tandem SPE [8,9], which are tedious, time-consuming, require a large number of reagents and chemicals, and generally offer a low recovery. Subsequently, many laboratories have also proposed other faster and more convenient strategies. Feng et al. (2022) established a method of magnetic solid-phase extraction (MSPE) based on novel magnetic covalent organic polymers for the detection of HAAs in fish and meat products [10]. Chevolleau et al. (2020) developed a method that combines SALLME extraction and SPE purification for the analysis of HAAs in cooked beef [11]. Recent studies (Hsiao et al., 2017; Chiang et al., 2022; Lai et al., 2023) have developed the QuEChERS method using a combination of PSA, $MgSO_4$ and C18EC as the clean-up adsorbents to analyze HAAs from meat products, and proved that the QuEChERS method is a simple, rapid and convenient sample pretreatment method with broad applicability [12–14].

These outstanding superiorities of QuEChERS make it a popular technique in the analysis of contaminants in complex food samples, including HAAs. It includes two processes, extraction and clean-up [15]. The key step of the QuEChERS method lies in its clean-up process, and the clean-up adsorbent is an essential factor. The commonly used clean-up adsorbents include C18, graphitized carbon black (GCB) and primary secondary amine (PSA) [16]. PSA with primary and secondary amino groups could form strong hydrogen bonds with some matrix components. Thus, it exhibits the function of removing organic acids, some sugars and fatty acids. C18 has a strong adsorption capacity for non-polar interfering substances and GCB is mainly applied to remove pigments. However, when C18, PSA or GCB is used alone, the clean-up performance is not satisfactory for the treatment of complex matrix samples, and low sensitivity and recovery may occur. Therefore, these traditional adsorbents usually need to be used in combination, and the combination of multiple clean-up adsorbents made experimental procedures more tedious [17]. In order to solve the problems, the development of more efficient clean-up adsorbents is imperative. Moreover, high-speed centrifugation is inevitably used to separate the clean-up adsorbent from extracting solution after purification in the conventional QuEChERS approach, which makes the pretreatment process cumbersome and time-consuming.

Recently, multi-walled carbon nanotubes (MWCNTs) have attracted substantial interest in sample pretreatment, due to their unique hollow tubular structures, high surface area, and π-π stacking interactions; these features enable them to exhibit exceptional capabilities of matrix purification [18]. Nevertheless, when MWCNTs are served as clean-up adsorbents, the major challenge is their separation and recovery from water-based solutions. The magnetic functionalization of MWCNTs with Fe_3O_4 magnetic nanoparticles could achieve rapid separation and collection of Fe_3O_4-MWCNTs via a permanent magnet, which endows clean-up adsorbents with new capabilities, and preserves the inherent advantages of MWCNTs [19]. Compared with the classical QuEChERS process, magnetic separation needs no additional centrifugal devices, making the QuEChERS procedure more convenient and time-saving.

Braised sauce beef, namely "Jiang Niu Rou", is traditionally served with ready-to-eat meat products. It is popular in some Asian countries, including China. Although braised sauce beef is processed at a relatively low temperature, and is often repeatedly boiled for a long time together with "Lao Lu", namely the marinating juice [20]. Conventionally, the marinating juice is often used for several or dozens of cooking cycles, for even several months or years, which offers good flavor to braised sauce beef. However, the amount of precursors in the recycled marinating juice, such as glucose, creatine and free amino acids, increases significantly with increasing cooking cycles, facilitating the formation of various hazardous substances, such as HAAs [21]. Zhou et al. [8] discovered that several HAAs were found in braised sauce beef, and the content of HAAs increased with repeated cooking times of beef. Therefore, more attention should be paid to the monitoring and measuring of HAAs in braised sauce beef.

This study aimed to describe a modified QuEChERS technique based on Fe_3O_4-MWCNTs coupled with HPLC to simultaneously determine 2-amino-1-methylimidazo [4,5-b]quinoline (IQ [4,5-b]), 1-methyl-9*H*-pyrido [3,4-b]indole (Harman), 2-amino-3-methylimidazo [4,5-f]quinoline (IQ), 9*H*-pyrido [3,4-b]indole (Norharman) and 2-amino-5-phenylpyridine (Phe-P-1) in braised sauce beef. The experimental parameters in the modified QuEChERS process were optimized. The proposed technique was validated on the basis of its analytical performance, and finally applied to analyze real samples of braised sauce beef. This study not only offers a novel way for analyzing HAAs, but also has practical application values for routine monitoring in food safety.

2. Materials and Methods

2.1. Reagents and Materials

HAA standards, including 9*H*-pyrido [3,4-b]indole (Norharman), 2-amino-3-methylimidazo [4,5-f]quinoline (IQ), 1-methyl-9*H*-pyrido [3,4-b]indole (Harman), 2-amino-5-phenylpyridine (Phe-P-1) and 2-amino-1-methylimidazo [4,5-b]quinoline (IQ [4,5-b]), were ordered from Toronto Research Chemicals (North York, Ontario, Canada); their chemical structures are presented in Figure S1. All of the standards had a purity > 98%. All HPLC-grade reagents used in the HPLC system, including acetonitrile (ACN), methanol, ammonium acetate and acetic acid, were ordered from Sigma-Aldrich (Shanghai, China). All analytical-grade reagents, including anhydrous sodium acetate (NaOAc), anhydrous magnesium sulfate ($MgSO_4$), sodium chloride (NaCl) and anhydrous sodium sulfate (Na_2SO_4), were supplied by Aladdin Biochemical Technology Co., Ltd. (Shanghai, China). C18, PSA, multi-walled carbon nanotubes (MWCNTs, purity > 98 wt%), -COOH functionalized MWCNTs (MWCNTs-COOH, 10–30 μm, purity > 98 wt%) and -OH functionalized MWCNTs (MWCNTs-OH, 10–30 μm, purity > 98 wt%) were provided by ANPEL Laboratory Technologies (Shanghai, China).

2.2. HPLC Conditions

An agilent 1290 HPLC system, which contains a photodiode-array detector (DAD) and a Zorbax XDB C18 column (4.6 mm × 250 mm, 5 μm), was used for separation. The solvent system containing A (acetonitrile) and B (10 mM ammonium acetate buffer solution, pH 4.5 adjusted with acetic acid) was pumped at 1.0 mL/min at 30 °C. A chromatographic gradient was applied and presented as follows: 0–5.0 min, 15% A; 5.0–5.1 min, with A increased from 15% to 20% gradually; 5.1–11 min, 20% A; 11.0–11.1 min, with A increased from 20% to 30% gradually; 11.1–23.0 min, 30% A. The injection volume was 20 μL. A wavelength of 258 nm was used. The typical chromatograms of the standard solution, spiked real sample and blank real sample are displayed in Figure S2. The retention time of each HAA was 5.85 min, 10.13 min, 14.42 min, 15.40 min and 19.50 min.

2.3. Preparation of Fe_3O_4-MWCNTs

Fe_3O_4-MWCNTs were prepared via a solvothermal method with slight modification [22]. Briefly, 100 mg of MWCNTs were dispersed in 50 mL of ethylene glycol, with bath ultrasonication for 120 min to form a homogeneous solution. Amounts of 0.675 g of $FeCl_3 \cdot 6H_2O$ and 0.5 g of polyethylene glycol were placed into the above black solution and sonicated for another 20 min. Subsequently, 1.8 g of NaOAc was added under magnetic stirring for 10 min. The obtained mixture was heated for 10 h at 200 °C in a 100 mL Teflon-lined stainless steel autoclave. After reaction, the black precipitate was collected by a permanent magnet, followed by washing several times with water and ethanol to remove all of the unreacted chemicals and impurities. Black solid materials were obtained after drying at 50 °C.

2.4. Sample Preparation

All braised sauce beef samples were collected from local deli counters, and thoroughly minced with a kitchen blender to obtain a homogeneous powder. An aliquot (2.5 g) of the braised sauce beef powder was added to a 50 mL centrifuge tube, which contained 7.5 mL of deionized water and 10 mL of extraction solvent (ACN containing 1% ammonium hydroxide), and vortexed violently for 1 min for extraction. Then, 2.0 g of $MgSO_4$ and 0.5 g of NaOAc were added, followed by vortexing for 60 s. After centrifugation (1 min, 4000× *g*), the acetonitrile layer was obtained for further purification.

For the clean-up procedure, the acetonitrile supernatant (6 mL), together with Fe_3O_4-MWCNTs (35 mg), were placed into another 10 mL centrifuge tube. After vortexing for 60 s, the clean-up agent was magnetically separated from the sample solution. After separation, a 2 mL aliquot of the purified extract was collected and blown with nitrogen to remove the solvent. The residue was redissolved in methanol before HPLC analysis.

For comparison, the traditional QuEChERS process with C18, PSA or the mixture of C18 and PSA (C18 + PSA) as clean-up adsorbents was additionally provided.

2.5. Gravimetric Measurement of Co-Extracts

There was 20 mL of acetonitrile extracts, with or without purification through the procedure of Section 2.4, taken into 50 mL glass flasks that were previously weighed. Then, the acetonitrile solvent was removed by blowing with nitrogen, and the residue was accurately weighed. The weight difference was utilized to evaluate the clean-up efficiency of matrix co-extracts using different clean-up adsorbents. All of the experiments were carried out in triplicate, and the results were presented as means $\pm$ standard deviations.

2.6. Method Validation

Method validation was conducted in accordance with SANTE guidelines (SANTE/12682/2019) [23] through the following parameters: linearity, limits of detection (LODs), limits of quantitation (LOQs), precision and accuracy.

For the evaluation of the linearity, the blank braised sauce beef samples were spiked with seven different concentrations, and then extracted using the procedure of Section 2.4. The HPLC peak areas were plotted against the respective HAA concentrations to construct the working curves. The linearity of each target compound was assessed according to the correlation coefficient (R). For target analytes, LOD and LOQ were experimentally determined through the HPLC analysis of serial dilutions of the mixed standard solution to reach signal-to-noise ratios (S/N) $\geq$ 3 and (S/N) $\geq$ 10, respectively [24]. Meanwhile, the LOQ corresponded to the minimum concentration of HAAs in braised sauce beef that met the criterion, with recoveries between 70% and 120% and RSD $\leq$ 20% [23,25]. The accuracy was tested with recovery experiments that were carried out at high, middle, and low concentrations (500, 300 and 100 ng/g), with six replicates for each level. The intra- and inter-day precisions were tested by analyzing blank spiked samples at three different levels on one day and three consecutive days, respectively. The RSDs of the intra- and inter-day tests represented the precision.

3. Results and Discussion

3.1. Characterization of Fe_3O_4-MWCNTs

In our study, Fe_3O_4-MWCNTs were obtained using the solvothermal method with no need for nitrogen protection, which simplified the preparation process. Moreover, Fe_3O_4-MWCNTs obtained by this method possessed more stable properties. Transmission electron microscopy (TEM, JEM2100, Tokyo, Japan) was employed to investigate the morphology of the Fe_3O_4-MWCNTs, and the TEM image is demonstrated in Figure S3. It could be observed from Figure S3 that Fe_3O_4 nanoparticles with nearly spherical shapes were scattered on the MWCNTs' surface, which indicated the successful preparation of Fe_3O_4-MWCNTs. Moreover, only fewer bonding sites of the surface of MWCNTs were occupied by Fe_3O_4 nanoparticles. Therefore, they had little effect on the adsorption performance of MWCNTs.

3.2. Optimization of the Extraction Process

3.2.1. Selection of Extraction Solvent

Complete extraction of HAAs from the sample matrix plays an essential role in achieving acceptable recoveries. Therefore, an appropriate optimized extraction solvent is necessary. For the QuEChERS application, ACN was the most commonly used extraction solvent, since it is able to extract many compounds with different polarities and lower amounts of proteins, lipids and other lipophilic co-extractives [26]. Previous studies have revealed that adding ammonium hydroxide or acetic acid to ACN may receive a higher recovery of the analyte [27]. Hsiao et al. (2017) found that the acidic acetonitrile could not extract IQ [4,5-b] from meat when DAD was used as the detector [12]. Therefore, we compared the recoveries of HAAs extracted using ACN, ACN containing 1% acetic acid and ACN containing 1% ammonium hydroxide. The rest of the conditions of the experiment were as follows: water volume, 5 mL; extraction salt, $MgSO_4$/NaOAc; salt combination amount, 2.0 g/0.5 g; extraction time, 3 min; centrifugation time, 3 min; Fe_3O_4-MWCNTs amount, 35 mg/6 mL. As depicted in Figure 1A, adding ammonium hydroxide offered the best extraction efficiency for all of the analytes; most importantly, IQ [4,5-b] also achieved an acceptable recovery (81.6%) using DAD as the detector, which is an improvement over the research of Hsiao et al. (2017). Moreover, Yan et al. (2014) reported that HAAs with amphoteric character had high solubility in organic solvent under alkaline conditions, which may improve the recoveries of HAAs [28]. Given the overall results, ammonium hydroxide was selected as the solvent modifier of the extraction solvent for further extraction.

The content of ammonium hydroxide (NH_4OH) in acetonitrile also has an important influence on the extraction efficiency during pretreatment, since HAAs have amphoteric properties [26]. The influence of ammonium hydroxide addition (0.5%, 1%, 3% and 5%, *v/v*) on the recoveries of HAAs was evaluated. The rest of the conditions of the experiment were as follows: water volume, 5 mL; extraction salt, $MgSO_4$/NaOAc; salt combination amount, 2.0 g/0.5 g; extraction time, 3 min; centrifugation time, 3 min; Fe_3O_4-MWCNTs amount, 35 mg/6 mL. As depicted in Figure 1B, all of the recoveries extracted using ACN containing 1% ammonium hydroxide were within acceptable limits (70%–120%), and were a little higher than that of 0.5% ammonium hydroxide. It can be noted from Figure 1 that IQ had a higher overall recovery than those of the other four HAAs, which may be attributed to the fact that IQ had the weakest interaction with the matrix and is easier to be extracted. In addition, this phenomenon may also be related to matrix effect. Continuous increases in ammonium hydroxide percentages would bring out unacceptable recovery changes (>120%) for IQ.

Thus, the extraction solvent selected was acetonitrile with 1% ammonium hydroxide.

3.2.2. Selection of the Volume of Water

Since the QuEChERS procedure was originally designed to analyze the fruit and vegetable samples with higher water content, it is commonly recommended that adding water to the samples, which contain a small amount of water, could improve the extraction efficiency [29]. This may be due to the fact that the added water would make the target analytes in the samples more accessible to ACN. The influence of the water volume was evaluated in our research by increasing the volume from 0 to 10 mL. The rest of the conditions of the experiment were as follows: extraction solvent, ACN containing 1% ammonium hydroxide; extraction salt, $MgSO_4$/NaOAc; salt combination amount, 2.0 g/0.5 g; extraction time, 3 min; centrifugation time, 3 min; Fe_3O_4-MWCNTs amount, 35 mg/6 mL. As depicted in Figure 1C, 7.5 mL was sufficient to achieve the best extraction efficiency. Whereas, no significant difference in the recoveries of HAAs was observed over 7.5 mL. As a consequence, 7.5 mL of water was finally selected for the subsequent process.

3.2.3. Selection of the Type of Extraction Salt

In a QuEChERS method, another essential factor focused on the selection of an appropriate combination of extraction salts, which could induce phase separation between the organic layer and water through a higher salting-out effect. Furthermore, extraction salt reduces the solubility of the analytes in water, and facilitates their partitioning into the organic phase, thus obtaining a higher extraction efficiency [30,31]. In this study, three groups of combined salts were investigated: (a) 2.0 g of $MgSO_4$ and 0.5 g of NaOAc; (b) 2.0 g of Na_2SO_4 and 0.5 g of NaCl; (c) 2.0 g of $MgSO_4$ and 0.5 g of NaCl. The rest of the conditions of the experiment were as follows: extraction solvent, ACN containing 1% ammonium hydroxide; water volume, 7.5 mL; extraction time, 3 min; centrifugation time, 3 min; Fe_3O_4-MWCNTs amount, 35 mg/6 mL. After adding the above salt combinations to the extract solution, the upper organic phase was collected and analyzed to compare the extraction recovery. As presented in Figure 1D, the salt combination of $MgSO_4$/NaOAc provided slightly higher recoveries than the other salt combinations, while no significant differences were observed. According to Hsiao et al. (2017), the combination of $MgSO_4$/NaOAc was frequently used in the QuEChERS procedure. As a result, we decided to choose the combination of $MgSO_4$/NaOAc as the extraction salt.

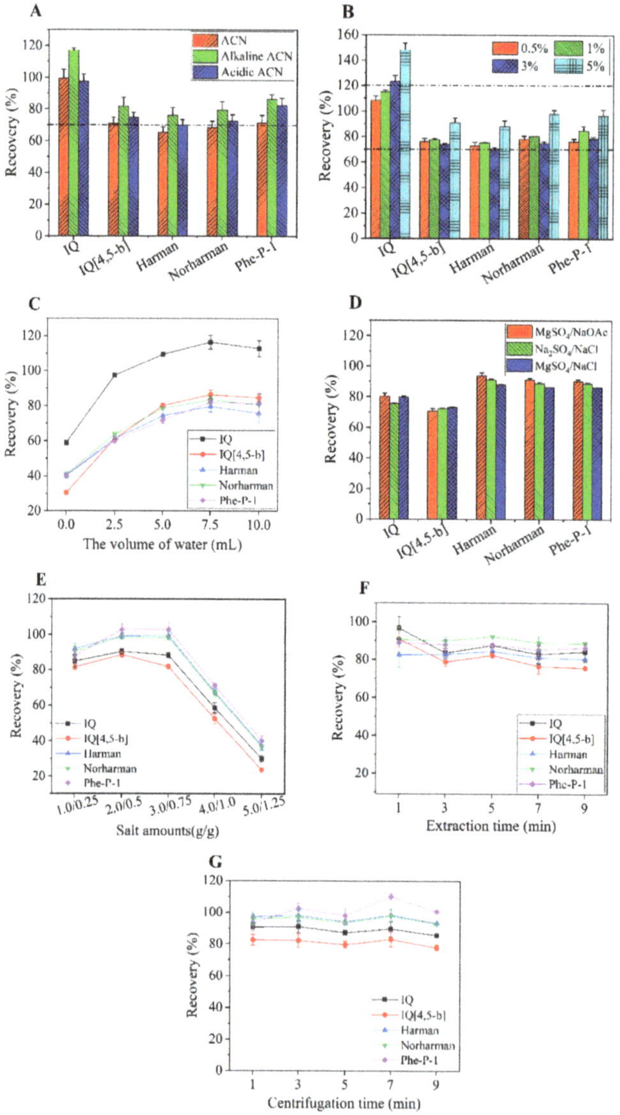

Figure 1. Effects of experimental conditions of the extraction process on the recovery. (**A**): acetonitrile, alkaline acetonitrile and acidic acetonitrile; (**B**): different percentages of ammonium hydroxide in acetonitrile; (**C**): volume of water; (**D**): type of extraction salt; (**E**): amount of $MgSO_4$/NaOAc; (**F**): extraction time; (**G**): centrifugation time. Spiked concentration, 300 ng/g.

Next, the influence of the amounts of the salt combination of $MgSO_4$/NaOAc on recoveries was further assessed, and the amounts were designed as follows: (1) 1.0 g/0.25 g; (2) 2.0 g/0.5 g; (3) 3.0 g/0.75 g; (4) 4.0 g/1.0 g; (5) 5.0 g/1.25 g. The mass ratio of $MgSO_4$ to NaOAc was fixed at 4:1. The rest of the conditions of the experiment were as follows: extraction solvent, ACN containing 1% ammonium hydroxide; water volume, 7.5 mL; extraction salt, $MgSO_4$/NaOAc; extraction time, 3 min; centrifugation time, 3 min; Fe_3O_4-MWCNTs amount, 35 mg/6 mL. As presented in Figure 1E, the recoveries of all five analytes reached the maximum when the amounts of the combination of $MgSO_4$/NaOAc were 2.0 g/0.5 g, while they rapidly decreased when the amounts were higher than 3.0 g/0.75 g. A portion of salts remained undissolved, and the volume of the acetonitrile layer was significantly reduced with a further increase in salt addition. We deduced that excessive addition of salts was not beneficial to phase separation, resulting in a marked decrease in the recoveries of target analytes. Thus, the selected amounts of $MgSO_4$/NaOAc were 2.0 g/0.5 g.

3.2.4. Selection of Extraction Time

The extraction time has a critical influence on the mass transfer process between the target analytes and extraction solvent, and thus affects the extraction recovery [32]. The vortex extraction time from 1 to 9 min was evaluated. The rest of the conditions of the experiment were as follows: extraction solvent, ACN containing 1% ammonium hydroxide; water volume, 7.5 mL; extraction salt, $MgSO_4$/NaOAc; salt combination amount, 2.0 g/0.5 g; centrifugation time, 3 min; Fe_3O_4-MWCNTs amount, 35 mg/6 mL. The results displayed in Figure 1F indicated that satisfactory recoveries could be achieved when the extraction time was 1 min. It was concluded that 1 min was sufficient to achieve thorough interaction between extraction solvent and the target analytes, while longer extraction times displayed no obvious change in the extraction yield of HAAs. Therefore, 1 min was employed for subsequent studies. Lai et al. (2023) reported that the shaking time after adding extraction solvent in the traditional QuEChERS method was 10 min [14]. Compared with this reference, the extraction time of this method was significantly shortened, which was conducive to shortening the whole sample pretreatment process. It can be attributed to the fact that large dosages of adsorbents (300 mg PSA, 900 mg $MgSO_4$ and 300 mg C18EC in 6 mL supernatant) in the traditional QuEChERS approach would take a long time to achieve complete extraction.

3.2.5. Selection of Centrifugation Time

The step of centrifugation is essential for completing the solid–liquid separation. Furthermore, centrifugation time also influences the volume of the upper organic layer [30]. Thus, an appropriate centrifugation time is vital for improving the extraction recovery. The influence of varying the centrifugation time from 1 to 9 min was evaluated. The rest of the conditions of the experiment were as follows: extraction solvent, ACN containing 1% ammonium hydroxide; water volume, 7.5 mL; extraction salt, $MgSO_4$/NaOAc; salt combination amount, 2.0 g/0.5 g; extraction time, 1 min; Fe_3O_4-MWCNTs amount, 35 mg/6 mL. As demonstrated in Figure 1G, satisfactory recoveries were obtained at 1 min, and the recoveries changed slightly when the centrifugation time was further prolonged. Therefore, 1 min was used in this experiment. The observed centrifugation time was significantly shorter than that reported in the literature of Lai et al. [14] (10 min), which was conducive to shortening the whole sample pretreatment process.

3.3. *Optimization of the Clean-Up Process*

3.3.1. Selection of the Adsorbent

Compared with traditional adsorbents PSA and C18, MWCNT materials have become a research hotspot in the field of clean-up adsorbents in recent years [17,31]. In this study, the clean-up performance of magnetic MWCNTs (Fe_3O_4-MWCNTs, Fe_3O_4-MWCNTs-COOH and Fe_3O_4-MWCNTs-OH) was compared with traditional adsorbents, including

C18 and PSA, as well as a mixture of PSA and C18 (PSA + C18). In the purification process, 7.5 mg of each type of magnetic MWCNT materials was added to 1 mL acetonitrile extracting solution of braised sauce beef, while 50 mg/mL were added for PSA or C18, and 100 mg/mL for the mixture of PSA and C18 (50 mg PSA and 50 mg C18). From the recovery results of Figure 2A, compared with the other clean-up adsorbents, the extracting solution purified using Fe_3O_4-MWCNTs displayed better or equivalent recoveries.

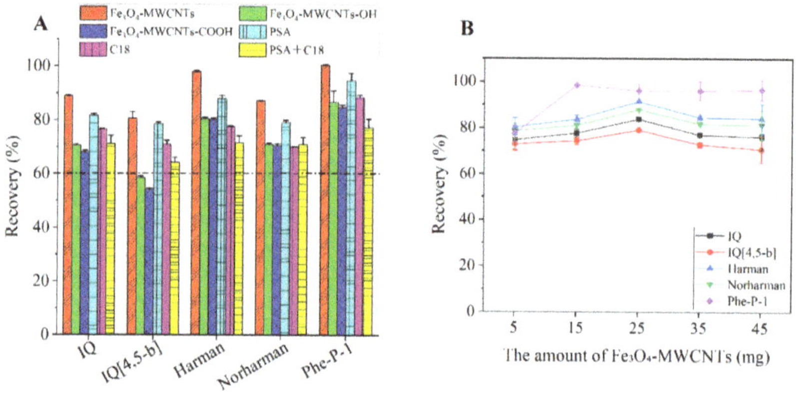

Figure 2. Effects of important factors of the clean-up process on the recovery. (**A**): type of adsorbent; (**B**): amount of Fe_3O_4-MWCNTs. Spiked concentration, 300 ng/g.

The clean-up efficiency in removing matrix co-extracts of the above-mentioned clean-up materials was further assessed using gravimetric analysis. As displayed in Figure 3, 7.3 mg of the remaining co-extracts was obtained with purification by Fe_3O_4-MWCNTs, while 16.5 mg was obtained without purification, suggesting that Fe_3O_4-MWCNTs was efficient in removing matrix co-extracts. Fe_3O_4-MWCNTs-COOH and Fe_3O_4-MWCNTs-OH provided superior removal abilities of matrix co-extracts than Fe_3O_4-MWCNTs, but the recovery of IQ [4,5-b] was less than 60% when they were used as the adsorbents. The reason for this phenomenon may be that the hydroxyl or carboxyl groups on the MWCNT surfaces strengthened the interaction of the matrix with the target compounds. In comparison, C18 possessed similar matrix removal efficiency to Fe_3O_4-MWCNTs, while its recoveries were lower than those of Fe_3O_4-MWCNTs. The matrix removal ability of PSA was the lowest. The combination of PSA and C18 displayed the largest clean-up efficiency, but their recoveries were lower than those of Fe_3O_4-MWCNTs. Besides, the combination of two clean-up materials made the pretreatment process troublesome, and the centrifugation process (3–10 min), which is needed to separate the adsorbents from the sample solution, was time-consuming.

Last but not least, when Fe_3O_4-MWCNTs were used as adsorbents, the solid–liquid separation could be completed within a few seconds by an external magnet without any additional centrifugation process. Moreover, compared with the combined effect of multiple adsorbents, only one adsorbent (Fe_3O_4-MWCNTs) used in the QuEChERS method made sample pretreatment easier and more time-saving, and facilitated high throughput analysis. Furthermore, the preparation of Fe_3O_4-MWCNTs was accomplished using a simple method and inexpensive MWCNTs. Considering their advantages of efficiency, speed, and simplicity as clean-up adsorbents, Fe_3O_4-MWCNTs could be used to analyze multiple targets in complex samples.

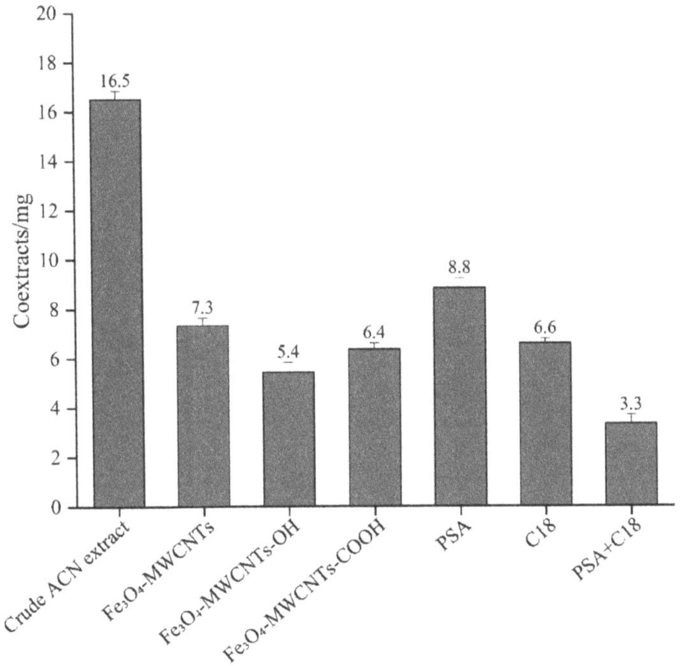

Figure 3. Amount of matrix co-extracts obtained through gravimetric measurement after purification with different clean-up materials (*n* = 3).

3.3.2. Selection of the Amount of Adsorbent

In order to obtain satisfactory recovery and good clean-up performance, the amount of Fe_3O_4-MWCNTs (5, 15, 25, 35 and 45 mg for 6 mL of the extract) was investigated. The rest of the conditions of the experiment were as follows: extraction solvent, ACN containing 1% ammonium hydroxide; water volume, 7.5 mL; extraction salt, $MgSO_4$/NaOAc; salt combination amount, 2.0 g/0.5 g; extraction time, 1 min; centrifugation time, 1 min. As described in Figure 2B, the recovery of Phe-P-1 was highest when the amount increased from 5 mg to 15 mg, but remained almost unchanged with a further amount increase. For the other four target compounds, all the recoveries were in satisfactory ranges (70–120%) with the increase of the amount of Fe_3O_4-MWCNTs, while, when it was over 25 mg, the recoveries decreased to some extent. Additionally, both the gravimetric test and the comparison of color of the purified solution were also performed to study the effect of Fe_3O_4-MWCNT amounts on the clean-up performance. The results demonstrated that the reduction of the co-extract weight became slow when the Fe_3O_4-MWCNT amounts exceeded 35 mg (Figure S4) and the final extract after purification with 35 mg of Fe_3O_4-MWCNTs was colorless and transparent (Figure S5). Given the overall results, the amount of Fe_3O_4-MWCNTs was selected as 35 mg/6 mL. Compared with the traditional QuEChERS method in Lai et al. (2023) in which the amounts of clean-up adsorbents were 300 mg PSA, 900 mg MgSO4 and 300 mg C18EC in 6 mL supernatant [14], the modified QuEChERS approach had less usage of adsorbent. Moreover, the cost of Fe_3O_4-MWCNTs is low (approximate $ 0.13 per 35 mg of adsorbent). Thus, the method possesses the advantage of having a low cost per sample.

3.4. Method Validation

The results of parameter validation are displayed in Table 1. All five target analytes had satisfactory linearity, with *R* values varying from 0.9983 to 0.9994. According to signal/noise ratios of 3 and 10, the LOQs and LODs for all five HAAs were 4.2–12.0 ng/g and 3.0–4.2 ng/g, respectively. The obtained LOQ value was lower than the prescribed maximum residue limit (MRL) recommended by European Union legislation, which sets the value at 50 μg/kg [33].

Table 1. Analytical performances of the method.

Analytes	Linear Range (μg/g)	*R*	Recovery ± RSD (%, *n* = 6)			Inter-Day RSD (%)	LOQs (ng/g)	LODs (ng/g)
			Low (100 ng/g)	Middle (300 ng/g)	High (500 ng/g)			
IQ	0.012–12.0	0.9989	82.8 ± 4.6	82.1 ± 3.2	82.3 ± 3.1	3.9	4.2	3.0
IQ [4,5-b]	0.012–12.0	0.9994	78.6 ± 2.9	78.8 ± 4.4	78.5 ± 2.5	4.6	12.0	4.2
Harman	0.012–12.0	0.9986	90.9 ± 3.8	89.6 ± 3.5	89.6 ± 3.3	4.1	4.2	3.0
Norharman	0.012–12.0	0.9991	89.1 ± 3.5	88.1 ± 3.4	88.2 ± 3.3	4.4	4.2	3.0
Phe-P-1	0.012–12.0	0.9983	103.2 ± 2.8	94.9 ± 0.6	101.6 ± 4.4	3.5	12.0	4.2

The method accuracy and precision were evaluated in terms of recoveries and the RSDs of the intra- and inter-day tests. According to the experimental results, the recoveries of the three concentration levels ranged from 78.5% to 103.2%. The intra-day and inter-day RSDs ranged between 0.6–4.6% and 3.5–4.6%, respectively. The analytical results of accuracy and precision were within the validation criteria (recovery between 70 and 120% with RSD ≤ 20%) [23]. Compared with previous literature, Hsiao et al. used LC–MS-MS analysis and a QuEChERS method with a mixture of PSA and C18 as adsorbents to extract HAAs from meat, and achieved a recovery of 58.9–117.4% with RSD less than 25.68% [12]. Jinap et al. conducted HPLC-DAD analysis on beef and chicken satay, employing Oasis MCX cartridge purification; they obtained recoveries of 43–92% and 49–98%, respectively [34]. The obtained recovery and RSD values were higher than, or comparable to, those achieved by the reported methods in the aforementioned literature, implying that the modified QuEChERS technique possessed excellent accuracy and precision.

3.5. Comparisons with Other Studies

The offered technique in this study was further compared with two reported techniques, namely, a frequently used technique (SPE) and a popular technique in recent years (QuEChERS). From the reported reference data in Table 2, the lowest recovery of the present method was obtained as 78.5%, compared to 28.9%, 43.0%, 49.0%, 56.2%, 52.3%, 58.9% and 42.5% as the corresponding values from other reported references. The precision RSD (<4.6%) obtained in this study was lower than those reported in all references in Table 2. This shows that this method displayed better or equivalent accuracy and precision. The LOD for the present method conforming to the criteria was slightly lower than those obtained with MS detector, but was acceptable. Meanwhile, the total sample pretreatment time for one sample required approximately 4 min, much shorter than that of the traditional QuEChERS technique using a mixture of C18 and PSA as adsorbents, and SPE methods. This meant that when Fe_3O_4-MWCNTs were used as the clean-up adsorbent, shorter extraction and centrifugation times in the extraction process coupled with magnetic separation in the clean-up process simplified the sample pretreatment process and made the separation more time-saving. In addition, the current method required the minimum usage of clean-up adsorbent compared with the traditional QuEChERS approach in [12]. The cost of Fe_3O_4-MWCNTs is low (approximate $ 0.13 per 35 mg of adsorbent). The comparison further proves that the modified QuEChERS method of applying Fe_3O_4-MWCNTs to extract HAAs possessed obvious superiorities in high precision and accuracy, speed, time saving and low cost per sample.

Table 2. Comparison of the present method with other methods.

Instrument	Extraction Method	Food Samples	Recovery (%)	RSD (%)	LOD	Sample Pretreatment Time (min)	Amounts of Adsorbent (mg/6 mL)	Ref.
UHPLC–APCI–MS/MS	SALLME [1] extraction/SPE purification	Cooked beef	60.7–108.5	<25.1	1–5 pg/μL	>100	-	[11]
LC–MS-MS	QuEChERS	Meat	58.9–117.4	<25.6	0.003–0.05 ng/mL	32	300 mg PSA + 900 mg $MgSO_4$ + 300 mg C18EC	[12]
HPLC-DAD	SPE [2]	Chicken/beef	43.0–92.049.0–98.0	-	0.3529–1.549 ng/g	>180	-	[34]
UPLC–MS/MS	Extrelut-SPE	Roasted beef patties	56.2–106.1	<11.2	0.008–0.053 ng/g	68.6	-	[35]
UHPSFC–MS/MS	SPE	Grilled meals/fish	52.3–97.5	<6.0	0.01–0.05 μg/kg	-	-	[36]
UHPLC–MS/MS	Acetonitrile/SPE	Meat products	42.5–99	<12.2	0.007–0.202 ng/g	>50	-	[37]
HPLC-DAD	SPE	Cooked meatballs	68.9–87.8	-	0.04–1.40 ng/g	>60	-	[38]
HPLC-DAD	Modified QuEChERS	Braised sauce beef	78.5–103.2	<4.6	3.0–4.2 ng/g	4	35	This study

[1] Salting-out liquid–liquid microextraction; [2] solid-phase extraction.

3.6. Analysis of Real Samples

Seventeen real samples of braised sauce beef (numbered Beef 1 to Beef 17) were obtained from different deli counters and analyzed using the above proposed method. As can be seen from Table S1, three analyzed HAAs (IQ, Harman and Norharman) were detected in Beef 1, Beef 8, Beef 13 and Beef 14. This was in accordance with the study of Yao et al. [39], who reported that IQ, Harman and Norharman were also detected from braised sauce beef; however, all of their concentrations were below the LOQ in our study. Only IQ was detected in Beef 5 at a concentration below the LOQ. Both Phe-P-1 and IQ [4,5-b] were not detected during our research, and none of the analyzed HAAs were detected in the other 12 samples.

The differences in HAAs in both amount and type among the different samples may be related to multiple factors. Lan et al. [40] reported that the addition of soy sauce and sugar in the marinating juice could facilitate the formation of HAAs in marinated pork, since sugar and soy sauce rich in amino acids were important precursors. Furthermore, according to Yao et al. [39], the HAA amounts in the chicken braised in soup cycled 20 times was approximately 7 times higher than that braised in soup cycled 5 times; moreover, the HAA content increased linearly with the cooking cycles of the marinating juice. Therefore, to clarify these differences, further research needs to be carried out in two aspects: (a) the analysis of HAAs in the marinating juice from different deli counters; (b) the factors affecting the formation of HAAs, including cooking method, cooking cycles of the marinating juice, and the type and quantity of seasonings and spices added to the marinating juice. Relevant studies are ongoing.

4. Conclusions

This report developed a modified QuEChERS technique based on Fe_3O_4-MWCNTs to determine HAAs in braised sauce beef samples combined with HPLC-DAD detection. Firstly, the Fe_3O_4-MWCNT materials were successfully prepared with a simple solvothermal method. Then, the conditions of the extraction process were optimized to comprise 10 mL of ACN containing 1% ammonium hydroxide, 7.5 mL of water, 2.0 g of $MgSO_4$ and 0.5 g of NaOAc, 1 min of vortex extraction time, and 1 min of centrifugation time. Then, a comparison with traditional adsorbents was carried out, and Fe_3O_4-MWCNTs displayed better or equivalent clean-up efficiency and experimental recoveries than PSA and C18. Considering the effect of the amount of Fe_3O_4-MWCNTs on clean-up performance and recoveries, 35 mg/6 mL was employed for optimal usage. The sample pretreatment for one sample was completed in about 4 min. Magnetic separation simplified the pre-processing operation, and made the separation more time-saving. Reduced usage of Fe_3O_4-MWCNTs provided the method another advantage of having a low cost per sample. Under optimized conditions, this method had good accuracy and precision, and acceptable LOQs and LODs. The recoveries were between 78.5% and 103.2%, and the precision was lower than 4.6%. The LODs of the method were 3.0–4.2 ng/g. Finally, the method was also applied to analyze 17 real braised sauce beef samples, and three targets (IQ, Harman and Norharman) were detected under the LOQ in 5 samples. In general, the proposed method, with its simplicity, time savings and low cost per sample, could be suitable for the routine monitoring of HAAs in braised sauce beef samples; moreover, Fe_3O_4-MWCNTs could be used as an alternative adsorbent in the QuEChERS method.

Supplementary Materials: The following supporting information can be downloaded at: https://www.mdpi.com/article/10.3390/foods12010138/s1, Figure S1: Chemical structures of the studied HAAs; Figure S2: Chromatograms of the mixed standard solution (A), the spiked braised sauce beef extract (B) and the blank braised sauce beef extract (C) prepared with the proposed QuEChERS method. Peak assignments: (1) IQ, (2) IQ [4,5-b], (3) Harman, (4) Norharman, (5) Phe-P-1; Spiked concentrations, 1.0 μg/g for IQ, Harman and Norharman, 0.5 μg/g for IQ [4,5-b] and Phe-P-1. Figure S3: TEM image of Fe_3O_4-MWCNTs; Figure S4: Amount of co-extracted matrix obtained through gravimetric analysis after purification with different amounts of Fe_3O_4-MWCNTs in the QuEChERS method ($n = 3$); Figure S5: Photographs of braised sauce beef extract after purification with different amounts of Fe_3O_4-MWCNTs; Table S1: HAAs measured in analyzed braised sauce beef samples.

Author Contributions: Conceptualization, M.L. and Y.B.; data curation, P.W.; formal analysis, P.W. and M.L.; funding acquisition, M.L. and Y.B.; investigation, M.L. and H.W.; methodology, P.W.; project administration, M.L. and X.Z.; resources, M.L.; software, H.W. and K.L.; supervision, M.L. and Y.B.; validation, M.L., P.W. and X.Z.; writing—original draft preparation, M.L.; writing—review and editing, M.L. and K.L. All authors have read and agreed to the published version of the manuscript.

Funding: This study was financed by the Key Scientific and Technological Project of Henan Province (grant number 212102310493, 222102310463), and the Doctoral Scientific Research Foundation of Zhengzhou University of Light Industry, (grant number 13501050065).

Data Availability Statement: Data are contained within the article and Supplementary Materials.

Conflicts of Interest: The authors declare no conflict of interest.

References

1. Shan, S.H.; Ma, Y.J.; Sun, C.L.; Guo, X.L.; Zheng, H.; Xu, X.B.; Qin, L.; Hua, J.N. A novel magnetic solid-phase extraction method for detection of 14 heterocyclic aromatic amines by UPLC-MS/MS in meat products. *Food Chem.* **2021**, *337*, 127630. [CrossRef] [PubMed]
2. Śnieżek, E.; Szumska, M.; Nowak, A.; Muzyka, R.; Janoszka, B. Application of high-performance liquid chromatography with fluorescence detection for non-polar heterocyclic aromatic amines and acridine derivatives determination in pork loin roasted in a roasting bag. *Foods* **2022**, *11*, 3385. [CrossRef] [PubMed]
3. Oz, F.; Kotan, G. Effects of different cooking methods and fat levels on the formation of heterocyclic aromatic amines in various fishes. *Food Control* **2016**, *67*, 216–224. [CrossRef]

4. Wang, W.; Dong, L.; Zhang, Y.; Yu, H.N.; Wang, S. Reduction of the heterocyclic amines in grilled beef patties through the combination of thermal food processing techniques without destroying the grilling quality characteristics. *Foods* **2021**, *10*, 1490. [CrossRef]
5. Linghu, Z.; Karim, F.; Taghvaei, M.; Smith, J.S. Determination of heterocyclic amines in meat matrices using enhanced matrix removal-lipid extraction and liquid chromatography-tandem mass spectrometry. *J. Food Sci.* **2019**, *84*, 1992–2002. [CrossRef]
6. Yue, X.Y.; Li, Y.; Xu, S.; Li, J.G.; Li, M.; Jiang, L.Y.; Jie, M.S.; Bai, Y.H. A portable smartphone-assisted ratiometric fluorescence sensor for intelligent and visual detection of malachite green. *Food Chem.* **2022**, *371*, 131164. [CrossRef]
7. Canales, R.; Mariño-Repizo, L.; Reta, M.; Cerutti, S. Multi-response optimization of a green solid-phase extraction for the analysis of heterocyclic aromatic amines in environmental samples. *Anal. Methods* **2020**, *12*, 1504. [CrossRef]
8. Zhou, Y.J.; Zhang, Y.X.; Dong, X.W. Determination of heterocyclic amines in braised sauce beef and the effects of different cooking conditions on the formation of heterocyclic amines. *J. Sci. Food Agric.* **2022**, *102*, 617–627. [CrossRef]
9. Yao, G.J.; Zhou, Y.J.; Li, Z.P.; Ma, Q.S. Heterocyclic aromatic amines in roasted chicken: Formation and prediction based on heating temperature and time. *Food Chem.* **2023**, *405*, 134822. [CrossRef]
10. Feng, Y.M.; Xu, Y.; Li, W.R.; Chen, S.S.; Su, Z.Q.; Xi, L.Y.; Li, G.L. Improved enrichment and analysis of heterocyclic aromatic amines in thermally processed foods by magnetic solid phase extraction combined with HPLC-MS/MS. *Food Control* **2022**, *137*, 108929. [CrossRef]
11. Chevolleau, S.; Bouville, A.; Debrauwer, L. Development and validation of a modified QuEChERS protocol coupled to UHPLC-APCI-MS/MS for the simple and rapid quantification of 16 heterocyclic aromatic amines in cooked beef. *Food Chem.* **2020**, *316*, 126327. [CrossRef] [PubMed]
12. Hsiao, H.Y.; Chen, B.H.; Kao, T.H. Analysis of heterocyclic amines in meat by the quick, easy, cheap, effective, rugged, and safe method coupled with LC-DAD-MS-MS. *J. Agric. Food Chem.* **2017**, *65*, 9360–9368. [CrossRef] [PubMed]
13. Chiang, C.F.; Liao, P.L.; Hsu, K.C.; Shen, J.Y.; Lin, J.L.; Yang, D.J. Establishment of optimal QuEChERS conditions of various food matrices for rapid measurement of heterocyclic amines in various foods. *Food Chem.* **2022**, *380*, 132184. [CrossRef] [PubMed]
14. Lai, Y.W.; Lee, Y.T.; Cao, H.; Zhang, H.L.; Chen, B.H. Extraction of heterocyclic amines and polycyclic aromatic hydrocarbons from pork jerky and the effect of flavoring on formation and inhibition. *Food Chem.* **2023**, *402*, 134291. [CrossRef]
15. García, N.G.; Zinola, G.; Macías, G.; Cesio, M.V.; Heinzen, H. Straightforward sample preparation method for the analysis of pesticide residues in ginger (*Zingiber officinale* Rosc.) using gas chromatography-tandem mass spectrometry. *Microchem. J.* **2022**, *78*, 107349. [CrossRef]
16. Ji, B.C.; Zhao, W.H.; Xu, X.; Han, Y.; Jie, M.S.; Xu, G.G.; Bai, Y.H. Development of a modified quick, easy, cheap, effective, rugged, and safe method based on melamine sponge for multi-residue analysis of veterinary drugs in milks by ultra-performance liquid chromatography tandem mass spectrometry. *J. Chromatogr. A* **2021**, *1651*, 462333. [CrossRef]
17. Wang, J.; Duan, H.L.; Fan, L.; Zhang, J.; Zhang, Z.Q. A magnetic fluorinated multi-walled carbon nanotubes-based QuEChERS method for organophosphorus pesticide residues analysis in *Lycium ruthenicum* Murr. *Food Chem.* **2021**, *338*, 127805. [CrossRef]
18. Wang, S.; Li, M.; Li, X.Q.; Li, X.J.; Li, X.M.; Li, S.Q.; Zhang, Q.H.; Li, H.M. A functionalized carbon nanotube nanohybrids-based QuEChERS method for detection of pesticide residues in vegetables and fruits. *J. Chromatogr. A* **2020**, *1631*, 461526. [CrossRef]
19. Xu, X.; Xu, X.Y.; Han, M.; Qiu, S.T.; Hou, X. Development of a modified QuEChERS method based on magnetic multiwalled carbon nanotubes for the simultaneous determination of veterinary drugs, pesticides and mycotoxins in eggs by UPLC-MS/MS. *Food Chem.* **2019**, *276*, 419–426. [CrossRef]
20. Guo, X.Y.; Zhang, Y.W.; Qian, Y.; Peng, Z.Q. Effects of cooking cycle Times of marinating juice and reheating on the formation of cholesterol oxidation products and heterocyclic amines in marinated pig hock. *Foods* **2020**, *9*, 1104. [CrossRef]
21. Yao, Y.; Peng, Z.Q.; Wan, K.H.; Shao, B.; Shi, J.M.; Zhang, Y.W.; Wang, F.L.; Hui, T. Determination of heterocyclic amines in braised sauce beef. *Food Chem.* **2013**, *141*, 1847–1853. [CrossRef] [PubMed]
22. Lu, W.H.; Fu, S.C.; Sun, X.Z.; Liu, J.; Zhu, D.Y.; Li, J.H.; Chen, L.X. Magnetic solid-phase extraction using polydopamine-coated magnetic multiwalled carbon nanotube composites coupled with high performance liquid chromatography for the determination of chlorophenols. *Analyst* **2021**, *146*, 6252–6261. [CrossRef] [PubMed]
23. *SANTE/12682/2019*; SANTE Guidance Document on Analytical Quality Control and Method Validation Procedures for Pesticide Residues Analysis in Food and Feed. European Commission Directorate General for Health and Food Safety Brussels: Brussels, Belgium, 2020.
24. Heo, H.M.; Jo, H.W.; Chang, H.R.; Moon, J.K. Development of Simultaneous Analytical Method for Imidazolinone Herbicides from Livestock Products by UHPLC-MSMS. *Foods* **2022**, *11*, 1781. [CrossRef] [PubMed]
25. Freitas, L.V.P.; Alves, L.M.G.; Sicupira, L.C.; Pinho, G.P.; Silvério, F.O. Determination of DDT in honey samples by liquid-liquid extraction with low-temperature purification (LLE-LTP) combined to HPLC-DAD. *Anal. Methods* **2021**, *13*, 1955–1964. [CrossRef]
26. Kemmerich, M.; Rizzetti, T.M.; Martins, M.L.; Prestes, O.D.; Adaime, M.B.; Zanella, R. Optimization by central composite design of a modified QuEChERS method for extraction of pesticide multiresidue in sweet pepper and analysis by ultra-high-performance liquid chromatography-tandem mass spectrometry. *Food Anal. Methods* **2015**, *8*, 728–739. [CrossRef]
27. Hu, M.F.; Qiu, J.S.; Zhang, H.; Fan, X.S.; Liu, K.F.; Zeng, D.Q.; Tan, H.H. Method development and validation of indaziflam and its five metabolites in soil, water, and fruits by modified QuEChERS and UHPLC-MS/MS. *J. Agric. Food Chem.* **2018**, *66*, 10300–10308. [CrossRef]

28. Yan, Y.; Zeng, M.M.; Zheng, Z.P.; He, Z.P.; Tao, G.J.; Zhang, S.; Gao, Y.H.; Chen, J. A novel one-step extraction method for simultaneously determining eleven polar heterocyclic aromatic amines in meat products by UHPLC-MS/MS. *Anal. Methods* **2014**, *6*, 6437–6444. [CrossRef]
29. Xing, J.L.; Zhang, Z.G.; Zheng, R.H.; Xu, X.R.; Mao, L.Y.; Lu, J.P.; Shen, J.; Dai, X.J.; Yang, Z.F. Simultaneous detection of seven alternaria toxins in mixed fruit puree by ultra-high-performance liquid chromatography-tandem mass spectrometry coupled with a modified QuEChERS. *Toxins* **2021**, *13*, 808. [CrossRef]
30. Tan, X.R.; Jin, Q.; Lu, J.W.; Zhao, B.; Gou, W.N.; Yang, R.; Fu, Y.; Xu, D.H.; Zhang, L. Development and validation of a modified QuEChERS method for the analysis of bisphenols in meats by UPLC-MS/MS. *Chromatographia* **2022**, *85*, 433–445. [CrossRef]
31. Wu, X.Q.; Tong, K.X.; Yu, C.Y.; Hou, S.; Xie, Y.Y.; Fan, C.L.; Chen, H.; Lu, M.L.; Wang, W.W. Development of a high-throughput screening analysis for 195 pesticides in raw milk by modified QuEChERS sample preparation and liquid chromatography quadrupole time-of-flight mass spectrometry. *Separations* **2022**, *9*, 98. [CrossRef]
32. Xu, C.; Liao, Y.Y.; Fang, C.Y.; Tsunoda, M.; Zhang, Y.X.; Song, Y.T.; Deng, S.M. Simultaneous analysis of ursolic acid and oleanolic acid in guava leaves using QuEChERS-based extraction followed by high-performance liquid chromatography. *J. Anal. Methods Chem.* **2017**, *2017*, 2984562. [CrossRef] [PubMed]
33. European Commission. Regulation (EC) No 396/2005 of the European Parliament and of the Council of February 23, on Maximum Residue Levels of Pesticides in or on Food and Feed of Plant and Animal Origin and Amending Council Directive 91/414/EEC, Official Journal of the European Union, Strasbourg. 2005. Available online: https://eur-lex.europa.eu/legal-content/EN/TXT/HTML/?uri=CELEX:32017R0978&from=EN (accessed on 16 January 2022).
34. Jinap, S.; Mohd-Mokhtar, M.S.; Farhadian, A.; Hasnol, N.D.S.; Jaafar, S.N.; Hajeb, P. Effects of varying degrees of doneness on the formation of Heterocyclic Aromatic Amines in chicken and beef satay. *Meat Sci.* **2013**, *94*, 202–207. [CrossRef] [PubMed]
35. Li, Y.; He, J.L.; Quan, W.; He, Z.Y.; Qin, F.; Tao, G.J.; Wang, Z.J.; Zeng, M.M.; Chen, J. Effects of polyphosphates and sodium chloride on heterocyclic amines in roasted beef patties as revealed by UPLC-MS/MS. *Food Chem.* **2020**, *326*, 127016. [CrossRef] [PubMed]
36. Gibis, M.; Kruwinnus, M.; Weiss, J. Impact of different pan-frying conditions on the formation of heterocyclic aromatic amines and sensory quality in fried bacon. *Food Chem.* **2015**, *168*, 383–389. [CrossRef]
37. Yan, Y.; Zhang, S.; Tao, G.J.; You, F.H.; Chen, J.; Zeng, M.M. Acetonitrile extraction coupled with UHPLC-MS/MS for the accurate quantification of 17 heterocyclic aromatic amines in meat products. *J. Chromatogr. B* **2017**, *1068–1069*, 173–179. [CrossRef]
38. Özdestan, Ö.; Kaçar, E.; Keşkekoğlu, H.; Üren, Ali. Development of a new extraction method for heterocyclic aromatic amines determination in cooked meatballs. *Food Anal. Methods* **2014**, *7*, 116–126. [CrossRef]
39. Yao, Y.; Peng, Z.Q.; Shao, B.; Wan, K.H.; Wang, F.L.; Zhang, Y.W.; Li, J.K.; Hui, T. Effects of frying and boiling on the formation of heterocyclic amines in braised chicken. *Poultry Sci.* **2013**, *92*, 3017–3025. [CrossRef]
40. Lan, C.M.; Chen, B.H. Effects of soy sauce and sugar on the formation of heterocyclic amines in marinated foods. *Food Chem. Toxicol.* **2002**, *40*, 989–1000. [CrossRef]

foods

MDPI

Article

Fluorescent Sensing of Ciprofloxacin and Chloramphenicol in Milk Samples via Inner Filter Effect and Photoinduced Electron Transfer Based on Nanosized Rod-Shaped Eu-MOF

Xiaoyue Yue [1], Chaoyun Wu [1], Zijun Zhou [2], Long Fu [1] and Yanhong Bai [1,3,*]

1 College of Food and Bioengineering, Zhengzhou University of Light Industry, Zhengzhou 450001, China
2 Henan Institute of Product Quality Supervision and Inspection, Zhengzhou 450047, China
3 Henan Key Laboratory of Cold Chain Food Quality and Safety Control, Zhengzhou 450001, China
* Correspondence: 2005067@zzuli.edu.cn

Abstract: Rapid, facile, and accurate detection of antibiotic residues is vital for practical applications. Herein, we designed a sensitive, visual, and rapid analytical method for sensitive detection of ciprofloxacin and chloramphenicol based on a nanosized rod-shaped Europium metal organic framework (Eu-MOF). The fluorescent Eu-MOF was firstly synthesized by a simple synthetic route at room temperature, which displays a red emission. The mechanisms of detecting ciprofloxacin and chloramphenicol were confirmed to be the inner filter effect (IFE) and photoinduced electron transfer (PET). Under the optimized experimental conditions, the detection limits of the developed method for ciprofloxacin and chloramphenicol detection were 0.0136 and 3.16 μM, respectively. Moreover, the sensor was effectively applied for quantitative determination of ciprofloxacin and chloramphenicol milk samples with satisfactory recoveries of 94.5–102% and 97–110%, respectively. This work developed a new method for rapid detection of ciprofloxacin and chloramphenicol residues. In addition, the established method has potential practical application value for on-site safety regulation on antibiotic residues in animal-derived food.

Keywords: fluorescent sensing; antibiotic detection; inner filter effect; photoinduced electron transfer; metal organic framework

1. Introduction

As people pay increasingly more attention to environmental protection and public food safety, detection of ultra-trace pollutants in nature carries great significance to the construction of ecological civilization and sustainable development. Thereinto, the problem of antibiotic residues recently became a social problem in most countries, which should be resolved. Antibiotics are powerful weapons against infectious bacteria, which have been widely used in animal husbandry to prevent pathema and promote growth. Undoubtedly, rational use of existing antibiotics has effectively increased the production and revenue of animal husbandry and aquaculture industries. Nevertheless, their unreasonable abuse and excessive residues have also lead to catastrophic pollution in animal-derived food as well as water and soil resources, which threatens human health [1]. Therefore, it is particularly important for us to establish rapid, sensitive, and selective methods to detect antibiotic residues [2]. Ciprofloxacin is the most used fluoroquinolone drug for the treatment of pneumonia, lower respiratory tract infections, and skin infections. However, ciprofloxacin has slow metabolism and degradation, whose abuse can lead to excessive ciprofloxacin residue in animal-derived foods [3]. Besides, ciprofloxacin residues can cause a series of problems, including allergic reactions, intestinal flora disorders, antibiotic resistance, posing a threat to human health [4]. Chloramphenicol is often widely used as a cheap and effective veterinary drug in the field of animal husbandry. Chloramphenicol can bind to the 70S ribosome of human mitochondria, thus inhibiting protein synthesis in human mitochondria and causing aplastic anemia and visual impairment in humans [5,6]. At present, the United States, the European Union, China, etc., stipulate that chloramphenicol should not be detected in animal-derived foods [5,7]. Therefore, it is our inevitable choice to establish an efficient and sensitive method for the detection of ciprofloxacin and chloramphenicol.

There are many well-established antibiotic testing technologies to choose from, such as capillary electrophoresis [8], high-performance liquid chromatography [9], ion mobility spectrometry, Raman spectroscopy [10], and liquid chromatography-tandem mass spectrometry [11,12]. However, most of these methods have the disadvantages of dependence on expensive instruments, complicated operations, and need for necessary sample pretreatment. Therefore, it is necessary to establish a new and simple antibiotic detection method [13].

Fluorescent sensors enjoy the superiorities of high sensitivity, easy accessibility, and low cost, which means great attraction in detection. Generally, fluorescent sensors need to use excellent fluorescent probes as the signal platform to achieve the sensitive detection of targets. Nowadays, there have been many attempts to construct fluorescent sensors using different kinds of fluorescent materials as much as possible, which mainly include fluorescent carbon dots (CDs), II–VI semiconductor quantum dots, and metal organic frameworks (MOFs) materials. Thereinto, as a promising new type of crystalline material, metal organic frameworks (MOFs) materials are widely used in research fields, such as biochemical catalysis, gas storage, bioimaging, drug delivery, and chemical sensing. With unique advantages in chemical sensing, such as narrow fluorescence emission, high porosity, structural and chemical tunability, large Stokes shift values, rapid response, high sensitivity and selectivity, luminescent europium metal-organic framework (Eu-MOF) has become a smart material, that has attracted intense attention in the past few decades [14,15]. However, most of these experiments are often based on a single transmission signal, which is very susceptible to environmental interference, resulting in unstable experimental signals, which is not conducive to multiple experiments. Compared to the single fluorescence-based method, dual emission fluorescence sensors can minimize these effects and achieve higher analysis accuracy by self-calibrating the two fluorescence intensities. Generally speaking, with perceptible color change, dual emission fluorescence sensors support quick visual recognition. Therefore, the use of MOF-based dual emission ratio fluorescence sensors enjoys more promising prospects [16]. Due to unknown factors, such as environment, instrumentation, etc., single-channel fluorescence sensors typically have lower reliability and selectivity [17]. Dual emission fluorescence sensors are receiving increasing attention due to their advantages of eliminating background noise and improving signal-to-noise ratios [18,19]. Dual emission fluorescence probes rely on changes in the intensity of two or more emission bands caused by the analyte, by this method, the results may be more accurate and sensitive [20].

The sensing mechanisms observed in luminescent metal-organic frameworks materials generally include ligand metal energy transfer (LMET), fluorescence (Förster) resonance energy transfer (FRET), photoinduced electron transfer (PET), antenna effect, inner filter effect (IFE), skeleton collapse and adsorption [21–23]. The photoelectron property of europium ions is a result of 4f electrons, including long fluorescence lifetime, and sharp and stable emission bands. In addition, due to the "antenna effect", electrons of the ligand are transferred to the europium ions, which increases the luminous intensity. These characteristics greatly expand the application scope and development potential of Ln-MOF. At present, the application of Ln-MOFs in temperature sensing, small molecule sensing, metal ion sensing, and anion sensing has been widely reported [24].

In this work, a rapid fluorescence detection method was established for the detection of ciprofloxacin and chloramphenicol. First, Eu-MOF with a red fluorescent emission was synthesized. Eu-MOF consists of Eu^{3+} as a metal node and 1,3,5-benzenetricarboxylic acid, which acts like a bridge, causing a strong red fluorescence under the stimulation of excitation light. Based on the inner filter effect and photoelectron-induced transfer, chloramphenicol can quench the fluorescence of Eu-MOF. Moreover, the fluorescence intensity of ciprofloxacin can also be increased due to photoelectron-induced transfer, and the emission peak will also be higher. Therefore, Eu-MOF can be used as single emission and dual emission fluorescence sensors to detect chloramphenicol and ciprofloxacin in food, respectively.

2. Experimental

2.1. Main Reagents and Materials

Eu $(NO_3)_3{\cdot}6H_2O$ and trimesic acid (H_3BTC) were purchased from Aladdin (Shanghai, China). Ciprofloxacin (CIP), chloramphenicol (CHL), erythromycin (ERY), florfenicol (FLO), gentamycin sulfate (GEN), kanamycin Sulfate (KAN), lincomycin hydrochloride (LIN), neomycin (NEO), streptomycin (STR), and thiamphenicol (THI) were from Sangon Biotech Co., Ltd. (Shanghai, China).

2.2. Experimental Instruments

The morphology was investigated by scanning electron microscopy using a JSM-7001F. Powder X-ray diffraction (XRD) spectra were recorded using an X-ray device (D8 ADVANCE, Bruker Corp, Karlsruhe, Germany). Fourier transform infrared (FT-IR) spectra were recorded on an Antaris II FTIR spectrophotometer (San Jose, CA, USA). The fluorescence spectrum and the UV-vis absorption spectrum were measured on a Hitachi F-7000 fluorescence spectrophotometer and a TU-1810 spectrophotometer (Beijing Purkinje General Instrument Co., Ltd., Beijing, China), respectively. Fluorescence spectra were obtained from a Hitachi F-7000 fluorescence spectrophotometer (Hiyachi, Tokyo, Japan). The steady-state/transient fluorescence spectra were measured a FLS-980 transient fluorescence spectrophotometer (Livingston, UK). HPLC measurements were performed on an Agilent 1290 Infinity II HPLC (Agilent, Palo Alto, CA, USA).

2.3. The Preparation of Eu-MOF

The preparation of Eu-MOF referred to the previous method reported by Li's group with some modificatione [25]. Specifically, 10 mM Eu $(NO_3)_3{\cdot}6H_2O$ was dissolved in deionized water and 10 mM trimesic acid was dissolved in absolute ethanol via ultrasonic vibration. Then, the metal precursor was slowly added to the organic ligand solution, and magnetically stirred at room temperature for one hour. After centrifugation at 10,000 rpm min^{-1} for 3 min, the precipitate was washed with water and absolute ethanol, and then, the final product was dried at 60 °C for 6 h.

2.4. Fluorescence Detection of Ciprofloxacin and Chloramphenicol

We dispersed 0.004 g of Eu-MOF powder in 10 mL of deionized water by ultrasonic. Then, different concentrations of ciprofloxacin (0.1–18 μM) and chloramphenicol (5–150 μM) were added to the above Eu-MOF solution, respectively. The volume ratio of ciprofloxacin to Eu-MOF solution was 4:1, and the volume ratio of chloramphenicol to Eu-MOF solution was 12:1. The corresponding solution was added to a 3.5 mL quartz fluorescent cuvette, and all fluorescence measurements were repeated three times at the excitation wavelength of 258 nm.

2.5. Determination of Ciprofloxacin and Chloramphenicol in Milk

The liquid milk samples were bought from a local supermarket and pretreated according to a national standard of China (GB/T 29692 2013) with some modification. We added 100 μL phosphoric acid and 4 mL acetonitrile to 2 g liquid milk. Then, it was stirred at medium speed for 5 min. Afterwards, the above solution was centrifuged at 10,000 rpm min^{-1} for 10 min, and then transferred the upper supernatant to a beaker for future usage. Five mL n-hexane was added to the supernatant, vortexed for 1 min and left to stand for 5 min. Transferred the residues to a pear-shaped flask. Next, 4 mL of acetonitrile was added into the residue and the extraction was repeated once. The supernatant was combined with the same portion of n-hexane, and the two extracts were combined and rotary evaporated at 50 °C. The residue was dissolved in 50 mL of deionized water, and filtered through a 0.22 μM organic filter membrane for HPLC determination and FL analysis.

3. Results and Discussions

3.1. Physical Characterization of Eu-MOF

To deeply understand the properties of the prepared Eu-MOF, SEM, FTIR, and XRD were used to characterize the morphology and structures of Eu-MOF. Figure 1A,B shows that Eu-MOF is a typical nanosized rod-shaped structure with good monodispersity. The illustration is a photo of Eu-MOF solution under ultraviolet light and f natural light, respectively. Eu-MOF is a white translucent suspension under natural light. When the aqueous dispersion of Eu-MOF is exposed to 254 nm ultraviolet light, bright red fluorescence can be easily observed. In addition, the composition of Eu-MOF was further confirmed by FT-IR spectroscopy. As shown in Figure 1C, the strong absorption at 3408 cm^{-1} is attributed to the -OH stretching vibration group of H_2O, indicating that water molecules act as solvents and reactants at the same time. The characteristic bands appearing in 1613–1560 cm^{-1} and 1436–1374 cm^{-1}, respectively, represent the symmetric and antisymmetric stretching vibration of -COO group of the ionizing ligand [26]. The absorption peak at 534 cm^{-1} is attributed to Eu-O stretching vibration, indicating that Eu ion interacts with -COO group [27].

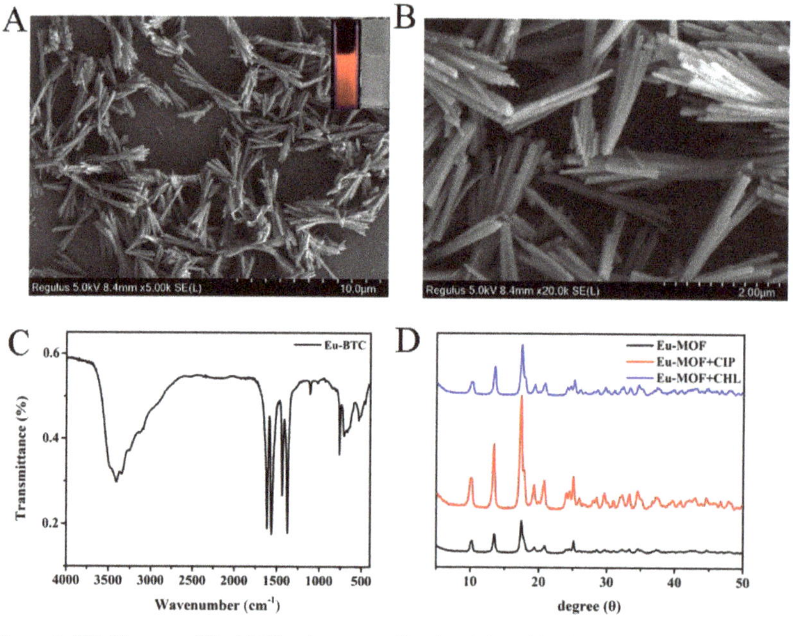

Figure 1. SEM images of Eu-MOF at low magnification (**A**) and high magnification (**B**); (**C**) FTIR spectra of Eu-MOF; (**D**) XRD spectra of Eu-MOF before and after the addition of CIP and CHL; Inset of (**A**): optical images of Eu-MOF solution under a UV lamp (left) and natural light (right).

3.2. Detection Mechanisms

Fluorescence-sensing mechanisms usually include structure collapse, energy absorption, and electron transfer [28,29]. To prove the mechanism for fluorescent detection of ciprofloxacin and chloramphenicol, we first measured the XRD spectra of Eu-MOF. As shown in Figure 1D, the crystal structure of Eu-MOF remains unchanged after the addition of ciprofloxacin or chloramphenicol, indicating that the Eu-MOF skeleton did not collapse, and it was ruled out that the skeleton collapse caused the fluorescence quenching of Eu-MOF. As can be seen from Figure 2A,B, the ultraviolet absorption peaks of ciprofloxacin and chloramphenicol almost completely overlap with the fluorescence excitation peaks of Eu-MOF, so the inner filter effect was one of the quenching mechanisms [1]. Later, to confirm the other detection mechanisms, the steady state transient fluorescence spectra of Eu-MOF were obtained in the absence or presence of ciprofloxacin and chloramphenicol. As shown in Figure 2C, the fluorescence lifetime of Eu-MOF does not change after the addition of chloramphenicol. However, the fluorescence lifetime of Eu-MOF increases to some extent after the addition of ciprofloxacin, which may be caused by the blue fluorescence of ciprofloxacin itself. In addition, the quenching time of Eu-MOF is less than 20 s, so the quenching mechanism may not be that antibiotics are adsorbed into the voids of Eu-MOF. Therefore, the inner filter effect was one of the mechanisms for ciprofloxacin and chloramphenicol detection based on fluorescent Eu-MOF.

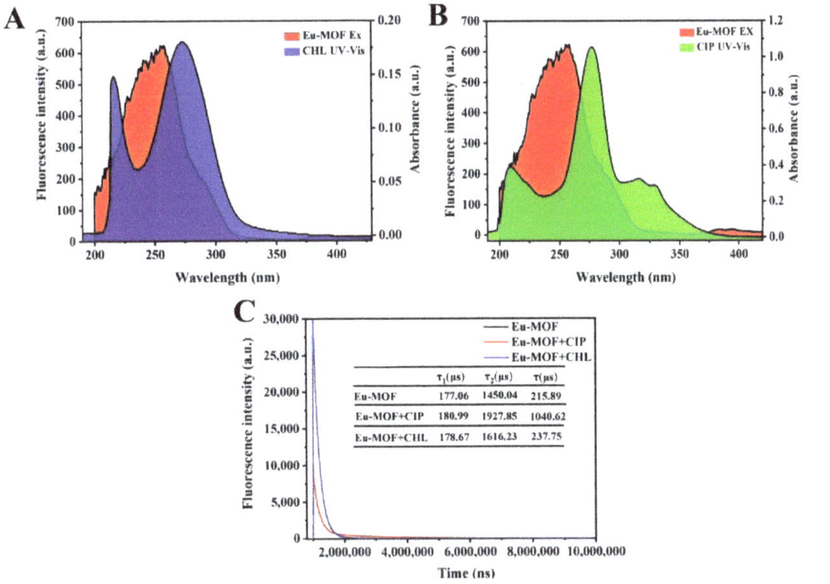

Figure 2. Fluorescence excitation spectra of Eu-MOF overlaid with the UV-vis absorption spectra of ciprofloxacin (**A**) and chloramphenicol (**B**); (**C**) fluorescence lifetime spectra of Eu-MOF before and after the addition of ciprofloxacin and chloramphenicol.

To analyze whether the detection mechanisms include photoelectron-induced transfer besides the inner filter effect, the highest occupied molecular orbital (HOMO) energy was calculated based on the ultraviolet photoelectron spectroscopy (UPS) of Eu-MOF. HOMO energy was determined by the incident light energy ($h\upsilon$ = 21.2 eV), onset binding energy (E_{onset}), and cutoff binding energy (E_{cutoff}) [30]. The formula is as follows:

$$E_{HOMO} = h\upsilon - (E_{cutoff} - E_{onset}) \quad (1)$$

Figure 3 shows the UPS spectrum of Eu-MOF (Figure 3A) and the UV-vis absorption spectra of Eu-MOF (Figure 3C). The abscissa is the binding energy relative to the Fermi energy (EF) of Au, which is defined by the energy of the electron relative to the vacuum level before excitation. Figure 3A shows the high binding energy cutoff (E_{cutoff}) of Eu-MOF. E_{cutoff} was calculated according to the linear extrapolation of the secondary electron yield to zero. Figure 3A shows that the high binding energy cutoff of Eu-MOF was 17.75 eV. Figure 3B shows the HOMO area of Eu-MOF. E_{cutoff} = 17.75 eV, E_{onset} = 3.39 eV, so E_{HOMO}

can be calculated as −6.84 eV based on the Equation (1). The band gap is 5.93 eV according to the UV absorption spectra of Eu-MOF. The energy of the lowest unoccupied molecular orbital E_{HOMO} = E_{LOMO}-band gap, so E_{LUMO} is −0.91 eV [21]. As shown in Figure 3D, E_{LUMO} of ciprofloxacin is −1.9 eV and E_{LUMO} of chloramphenicol = −3.42 eV. Under lower LUMO energy of the analyte, transition of excited state electrons is easier. Simply put, Eu-MOF can be regarded as a large "molecule". The energy levels of the valence band (VB) and conduction band (CB) can be described in a model similar to molecular orbitals (MOs). The CB energy level of MOF is higher than the lowest unoccupied molecular orbital (LOMO) of the analyte, which results in the driving force for electron transfer from MOF to the analyte, leading to fluorescence quenching [25,31]. Therefore, electrons may be transferred from Eu-MOF to ciprofloxacin and chloramphenicol under light-induced action. From Figure S1, we can clearly see that with the addition of ciprofloxacin to Eu-MOF solution, the fluorescence peak of ciprofloxacin at 449 nm gradually increases and stabilizes at 200 s, which also proves that the electrons of Eu-MOF transferred to ciprofloxacin, leading to the enhanced fluorescence of ciprofloxacin. Therefore, the detection mechanism should be ascribed to a joint result of the inner filter effect and photoelectron-induced transfer.

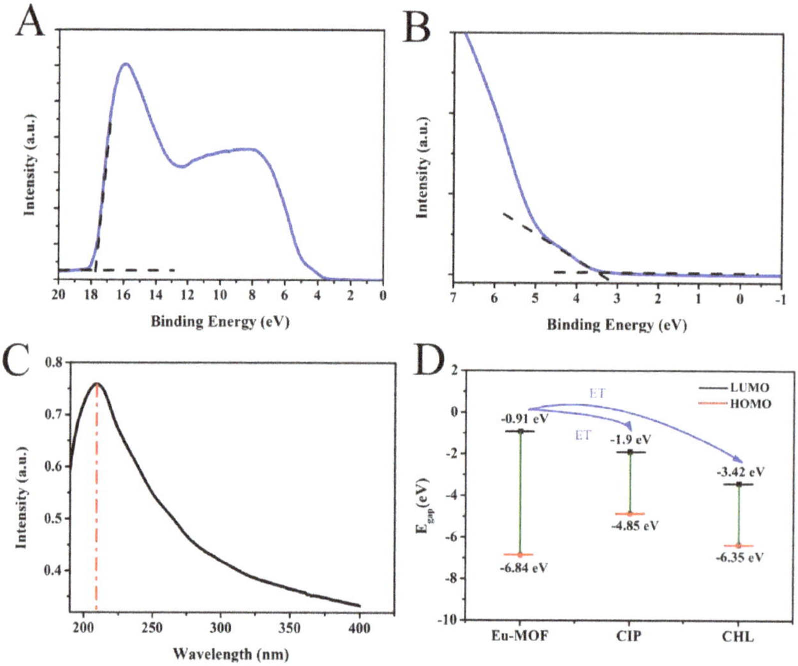

Figure 3. UPS spectra of Eu-MOF in (**A**) the second edge region and (**B**) the HOMO region; (**C**) UV absorption spectra of Eu-MOF; (**D**) HOMO-LUMO energy levels and molecular front orbital energy levels of Eu-MOF and CIP and CHL.

3.3. Optimization of Detection Conditions

The excitation wavelength of Eu-MOF was firstly optimized. As shown in Figure S1A, the fluorescence intensity of Eu-MOF increases with the increasing excitation wavelength. When the excitation light wavelength is 258 nm, the fluorescence intensity reaches the maximum. Afterwards, the fluorescence intensity of Eu-MOF gradually decreases with the increase of excitation wavelength. As the excitation wavelength changes, the emission wavelength does not shift, indicating that the Eu-MOF emission is excitation-independent emission [32,33]. Subsequently, the reaction time was investigated between the Eu-MOF sensor and antibiotics. As shown in Figure S1B,C, the reaction time of ciprofloxacin quenching Eu-MOF is less than 20 s. However, after the ciprofloxacin and Eu-MOF are mixed, its own fluorescence at 449 nm also increases until the fluorescence intensity stabilizes for 200 s. However, similar to ciprofloxacin, chloramphenicol can quench Eu-MOF in less than 20 s, allowing for instantaneous quenching of Eu-MOF fluorescence.

3.4. Fluorescence Detection of Ciprofloxacin and Chloramphenicol

Antenna effects are often observed in lanthanide complexes. Organic ligands absorb photons like antennas and generate tristates from single states through crossover between systems. Compared with free ions, the triplet makes lanthanide ions more sensitive and therefore has a brighter emission [16]. The energy gap between the singlet state and triplet state of ligand is key to effective emission, and the energy gap between the triplet state of the ligand and the excited state of Ln^{3+} ion is also strictly limited. Therefore, the luminescence of Eu-MOF is attributed to the energy transfer process from ligand to metal: The ligand absorbs the excited energy, and then transfers to the Eu^{3+} ion, resulting in Eu^{3+} luminescence [24]. Under the excitation of 258 nm, Eu-MOF displays three characteristic emission peaks located at 595, 618.8, and 690 nm, respectively. The maximum emission peak of Eu-MOF appears at 618.8 nm, which is the typical emission peak of Eu^{3+} [25].

Under the optimized conditions, the fluorescence sensor based on Eu-MOF was established to detect ciprofloxacin and chloramphenicol. As shown in Figure 4A, as the ciprofloxacin concentration gradually increases, the fluorescence of ciprofloxacin itself at 449 nm gradually increases. While the fluorescence intensity of Eu-MOF gradually decreases, and the fluorescent color of the detection solution shows a clear transition from red to blue (the inset of Figure 4A). Therefore, it is possible to use the ratio of fluorescence intensity at different positions to draw a standard curve. As exhibited in Figure 4B, there was a good relationship between the fluorescence intensity ratio ($F_{618.8}/F_{449}$) and the ciprofloxacin concentrations in a wide range of 0.1–18 μM with a lower detection limit of 0.0136 μM. Figure 4C shows the fluorescence spectrum of Eu-MOF after the addition of different concentrations of chloramphenicol. The fluorescence intensity of Eu-MOF gradually decreases with the increasing chloramphenicol concentration, showing obvious brightness transition from bright pink to dark pink. As shown in Figure 4D, there was a good relationship between the fluorescence intensity ratio (F_0/F) and the chloramphenicol concentrations in a wide range of 5–150 μM with a lower detection limit of 3.16 μM. Two linear ranges displayed in the range of 5–150 μM. The corresponding fluorescent intensity exhibited a linear response toward chloramphenicol concentrations within the range of 5–40 μM, which could be described by $F_0/F = 0.008\ c + 0.976$. However, the corresponding linear equation could be described by $F_0/F = 0.023\ c + 0.414$ in the linear range of 40–150 μM. In addition, compared with previously reported CIP and CHL assays (Table 1), the developed fluorescent sensor shows the proposed method displayed comparable or even superior analytical performance.

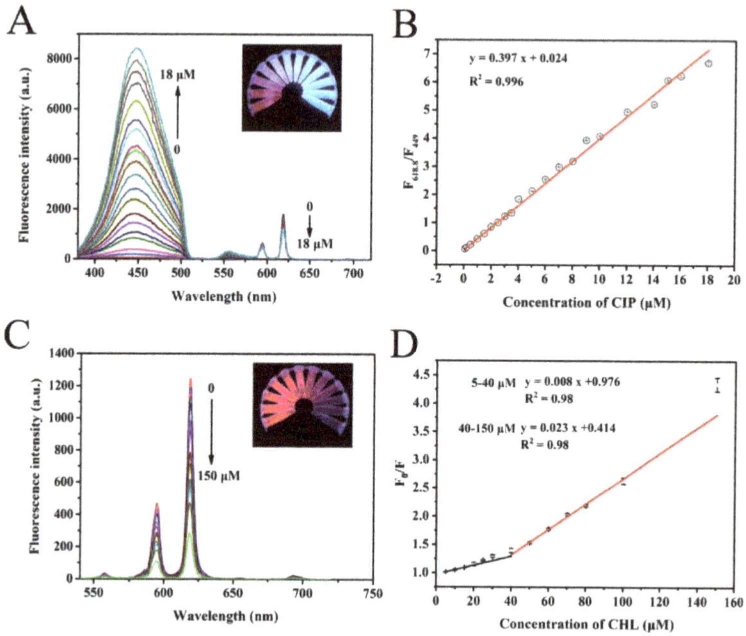

Figure 4. (**A**) Fluorescence spectra and (**B**) standard curve of Eu-MOF under the addition of different concentrations of ciprofloxacin; Inset of (**A**): optical images of fluorescent sensor after addition of different concentrations of ciprofloxacin in centrifuge tubes under a UV lamp at 254 nm; (**C**) fluorescence spectra and (**D**) standard curveof Eu-MOF under the addition of different concentrations of chloramphenicol; Inset of (**C**): optical images of fluorescent sensor after addition of different concentrations of chloramphenicol in centrifuge tubes under a UV lamp at 254 nm.

Table 1. The comparison of Eu-MOF with other methods for CIP and CHL detection.

Methods	Materials	Analytes	Linear Range	LOD	Ref.
Electrochemical	Graphene	Ciprofloxacin	0.1–100 μM	0.1 μM	[34]
Fluorescence	CdSe quantum dots	Ciprofloxacin	0–120 μM	0.6 μM	[35]
Electrochemical		Ciprofloxacin	10–80 μM	0.050 μM	[36]
Fluorescence	Eu^{3+} Doped in Sol-Gel Matrix	Ciprofloxacin	5.0×10^{-3}–1.0 μM	1.65×10^{-3} μM	[37]
Colorimetric	Triangular gold nanoplates	Chloramphenicol	0–2000 μM	5 μM	[38]
Electrochemical	Silver chloride/molybdenum disulfide	Chloramphenicol	4–531 μM	1.93 μM	[39]
Fluorescence	Eu-MOF	Ciprofloxacin Chloramphenicol	0.1–18 μM 5–150 μM	0.0136 μM 3.16 μM	This work

3.5. Anti-Interference Performance

To verify whether other antibiotics can affect the fluorescence of Eu-MOF, streptomycin, kanamycin sulfate, florfenicol, gentamicin, lincomycin hydrochloride, erythromycin, neomycin, and thiamphenicol were used to investigate the anti-interference performance of the developed sensor. As shown in Figure S2A, compared with the blank, the ratio of fluorescence intensity at 618.8 nm to fluorescence intensity at 449 nm is greater than 1 in the presence of only ciprofloxacin ($F_{618.8}/F_{449}$). When other antibiotics are added, $F_{618.8}/F_{449}$ is basically the same as that of blank, indicating that only ciprofloxacin has a strong emission peak at 449 nm. As shown in Figure S2B, only chloramphenicol can quench the fluorescence of Eu-MOF. Hence, the method for detecting ciprofloxacin and chloramphenicol established in this study has good anti-interference ability.

3.6. The Practical Application of Fluorescence Sensor Based on Eu-MOF in the Detection of Ciprofloxacin and Chloramphenicol

In order to explore the accuracy of the constructed fluorescence sensor in practical applications, pure milk purchased from the supermarket was used as the actual sample to detect the concentration of ciprofloxacin and chloramphenicol. As shown in Table 2, the results obtained by this method were consistent with those of high-performance liquid chromatography (HPLC). The results display that the recovery of the developed method for ciprofloxacin detection in milk was 94.5–102%, with a relative standard deviation of 3.6–9.1%. The recovery of the developed method for chloramphenicol detection in milk was 97–110%, with a relative standard deviation of 1.5–4.8%. Therefore, this method could be used for the detection of ciprofloxacin and chloramphenicol in milk.

Table 2. Detection of CIP and CHL in Spiked Milk Samples.

	Spiked Concentration (μM)	Eu-MOF	HPLC	Recoveries (%)	RSD (%)
CIP	2	2.001	2.152	100	3.6
	5	4.728	6.402	94.5	4.2
	10	10.196	12.362	102	9.1
CHL	10	9.704	12.106	97	4.8
	15	16.549	16.508	110	1.5
	20	20.898	20.133	104	3.6

4. Conclusions

In conclusion, a fluorescent sensor based on Eu-MOF was established for the rapid detection of ciprofloxacin and chloramphenicol. Firstly, Eu-MOF was successfully synthesized at room temperature. Then, sensitive, and selective detection of ciprofloxacin and chloramphenicol was achieved via the mechanism integrating the inner filter effect (IFE) and photoinduced electron transfer. The concentration ranges for ciprofloxacin and chloramphenicol detection were 0.1–18 μM and 5–150 μM, with low detection limits of 0.0136 μM and 3.16 μM, respectively. Furthermore, the established method was applied for ciprofloxacin and chloramphenicol detection in actual samples, showing satisfactory reproducibility, stability, and accuracy. Conclusively, the developed method is highly sensitive, selective, and capable of visual detection, which provides ideas for the future on-site visual semi-quantitative detection of antibiotics in food.

Supplementary Materials: The following supporting information can be downloaded at: https://www.mdpi.com/article/10.3390/foods11193138/s1, Figure S1: (A) Fluorescence emission spectra of Eu-MOF at different excitation wavelengths; (B) Fluorescence intensity of Eu-MOF at 449 nm and 618.8 nm after the addition of ciprofloxacin with time; (C) Fluorescence intensity of Eu-MOF at 618.8 nm after the addition of chloramphenicol with time; Figure S2: (A) The F449/F618.8 and (B) $(F_0 - F)/F_0$ at 618.8 nm of the Eu-MOF against various antibiotics.

Author Contributions: Conceptualization, X.Y. and Y.B.; methodology, C.W. and Z.Z.; software, C.W. and Z.Z.; validation, X.Y., C.W. and L.F.; formal analysis, X.Y. and Z.Z., investigation, C.W.; resources, X.Y.; data curation, C.W.; writing—original draft preparation, X.Y. and C.W.; writing—review and editing, X.Y and Y.B.; supervision, X.Y. and Y.B; project administration, X.Y.; funding acquisition, X.Y and Y.B. All authors have read and agreed to the published version of the manuscript.

Funding: This research was funded by the National Natural Science Foundation of China, (grant number 31801630), the major special project on public welfare of Henan Province (201300110100), and the Key Scientific and Technological Project of Henan Province (No: 222102310355).

Data Availability Statement: Data arecontained within the article.

Conflicts of Interest: The authors declare no conflict of interest.

References

1. Zhao, N.; Liu, J.; Yang, F.; Lv, S.; Wang, J.; Wang, S. Easy Green Construction of a Universal Sensing Platform Based on Crystalline Polyimide Covalent Organic Frameworks with Sensitive Fluorescence Response to Metal Ions and Antibiotics. *ACS Appl. Bio Mater.* **2020**, *4*, 995–1002. [CrossRef]
2. Jia, P.; Bu, T.; Sun, X.; Liu, Y.; Liu, J.; Wang, Q.; Shui, Y.; Guo, S.; Wang, L. A sensitive and selective approach for detection of tetracyclines using fluorescent molybdenum disulfide nanoplates. *Food Chem.* **2019**, *297*, 124969. [CrossRef]
3. Yuan, C.; He, Z.; Chen, Q.; Wang, X.; Zhai, C.; Zhu, M. Selective and efficacious photoelectrochemical detection of ciprofloxacin based on the self-assembly of 2D/2D g-C3N4/Ti3C2 composites. *Appl. Surf. Sci.* **2021**, *539*, 148241. [CrossRef]
4. Yuan, Y.; Zhang, F.; Wang, H.; Gao, L.; Wang, Z. A Sensor Based on Au Nanoparticles/Carbon Nitride/Graphene Composites for the Detection of Chloramphenicol and Ciprofloxacin. *ECS J. Solid State Sci. Technol.* **2018**, *7*, M201–M208. [CrossRef]
5. Wang, B.; Pang, M.; Zhao, X.; Xie, K.; Zhang, P.; Zhang, G.; Zhang, T.; Liu, X.; Dai, G. Development and comparison of liquid–liquid extraction and accelerated solvent extraction methods for quantitative analysis of chloramphenicol, thiamphenicol, florfenicol, and florfenicol amine in poultry eggs. *J. Mass Spectrom.* **2019**, *54*, 488–494. [CrossRef] [PubMed]
6. Xu, X.; Yang, Y.; Jin, H.; Pang, B.; Yang, R.; Yan, L.; Jiang, C.; Shao, D.; Shi, J. Fungal In Situ Assembly Gives Novel Properties to CdSxSe1−x Quantum Dots for Sensitive Label-Free Detection of Chloramphenicol. *ACS Sustain. Chem. Eng.* **2020**, *8*, 6806–6814. [CrossRef]
7. Xie, Y.; Zhao, M.; Hu, Q.; Cheng, Y.; Guo, Y.; Qian, H.; Yao, W. Selective detection of chloramphenicol in milk based on a molecularly imprinted polymer–surface-enhanced Ramanspectroscopic nanosensor. *J. Raman Spectrosc.* **2017**, *48*, 204–210. [CrossRef]
8. Hussain, A.; Alajmi, M.F.; Ali, I. Determination of chloramphenicol in biological matrices by solid-phase membrane micro-tip extraction and capillary electrophoresis. *Biomed. Chromatogr.* **2016**, *30*, 1935–1941. [CrossRef]
9. Liu, S.; Wu, X.-Z.; Gao, Z.-H.; Jiao, F. On-site solid phase extraction and HPLC determination of chloramphenicol in surface water and sewage. *Anal. Methods* **2013**, *5*, 1150–1154. [CrossRef]
10. Barveen, N.R.; Wang, T.-J.; Chang, Y.-H. Photochemical decoration of silver nanoparticles on silver vanadate nanorods as an efficient SERS probe for ultrasensitive detection of chloramphenicol residue in real samples. *Chemosphere* **2021**, *275*, 130115. [CrossRef]
11. Fedeniuk, R.W.; Mizuno, M.; Neiser, C.; O'Byrne, C. Development of LC–MS/MS methodology for the detection/determination and confirmation of chloramphenicol, chloramphenicol 3-O-β-d-glucuronide, florfenicol, florfenicol amine and thiamphenicol residues in bovine, equine and porcine liver. *J. Chromatogr. B* **2015**, *991*, 68–78. [CrossRef]
12. Vivekanandan, K.; Swamy, M.G.; Prasad, S.; Mukherjee, R. A simple method of isolation of chloramphenicol in honey and its estimation by liquid chromatography coupled to electrospray ionization tandem mass spectrometry. *Rapid Commun. Mass Spectrom.* **2005**, *19*, 3025–3030. [CrossRef] [PubMed]
13. Yang, H.-W.; Xu, P.; Ding, B.; Liu, Z.-Y.; Zhao, X.-J.; Yang, E.-C. A Highly Stable Luminescent Eu-MOF Exhibiting Efficient Response to Nitrofuran Antibiotics through the Inner Filter Effect and Photoinduced Electron Transfer. *Eur. J. Inorg. Chem.* **2019**, *2019*, 5077–5084. [CrossRef]
14. Zhao, Y.; Li, D. Lanthanide-functionalized metal–organic frameworks as ratiometric luminescent sensors. *J. Mater. Chem. C* **2020**, *8*, 12739–12754. [CrossRef]
15. Gan, Z.; Hu, X.; Xu, X.; Zhang, W.; Zou, X.; Shi, J.; Zheng, K.; Arslan, M. A portable test strip based on fluorescent europium-based metal–organic framework for rapid and visual detection of tetracycline in food samples. *Food Chem.* **2021**, *354*, 129501. [CrossRef] [PubMed]
16. Xing, K.; Fan, R.; Du, X.; Zheng, X.; Zhou, X.; Gai, S.; Wang, P.; Yang, Y. Dye-insertion dynamic breathing MOF as dual-emission platform for antibiotics detection and logic molecular operation. *Sens. Actuators B Chem.* **2019**, *288*, 307–315. [CrossRef]
17. Zhang, L.; Wang, Y.; Jia, L.; Bi, N.; Bie, H.; Chen, X.; Zhang, C.; Xu, J. Ultrasensitive and visual detection of tetracycline based on dual-recognition units constructed multicolor fluorescent nano-probe. *J. Hazard. Mater.* **2021**, *409*, 124935. [CrossRef]
18. Ti, M.; Li, Y.; Li, Z.; Zhao, D.; Wu, L.; Yuan, L.; He, Y. A ratiometric nanoprobe based on carboxylated graphitic carbon nitride nanosheets and Eu3+ for the detection of tetracyclines. *Analyst* **2021**, *146*, 1065–1073. [CrossRef] [PubMed]
19. Hu, J.; Yang, X.; Peng, Q.; Wang, F.; Zhu, Y.; Hu, X.; Zheng, B.; Du, J.; Xiao, D. A highly sensitive visual sensor for tetracycline in food samples by a double-signal response fluorescent nanohybrid. *Food Control* **2020**, *108*, 106832. [CrossRef]
20. Ye, Y.; Wu, T.; Jiang, X.; Cao, J.; Ling, X.; Mei, Q.; Chen, H.; Han, D.; Xu, J.J.; Shen, Y. Portable Smartphone-Based QDs for the Visual Onsite Monitoring of Fluoroquinolone Antibiotics in Actual Food and Environmental Samples. *ACS Appl. Mater. Interfaces* **2020**, *12*, 14552–14562. [CrossRef] [PubMed]
21. Yan, B. Luminescence response mode and chemical sensing mechanism for lanthanide-functionalized metal–organic framework hybrids. *Inorg. Chem. Front.* **2021**, *8*, 201–233. [CrossRef]
22. Zhao, Y.; Zeng, H.; Zhu, X.-W.; Lu, W.; Li, D. Metal–organic frameworks as photoluminescent biosensing platforms: Mechanisms and applications. *Chem. Soc. Rev.* **2021**, *50*, 4484–4513. [CrossRef] [PubMed]

23. Zhong, W.-B.; Li, R.-X.; Lv, J.; He, T.; Xu, M.-M.; Wang, B.; Xie, L.-H.; Li, J.-R. Two isomeric In(iii)-MOFs: Unexpected stability difference and selective fluorescence detection of fluoroquinolone antibiotics in water. *Inorg. Chem. Front.* **2020**, *7*, 1161–1171. [CrossRef]
24. Li, C.; Zhang, F.; Li, X.; Zhang, G.; Yang, Y. A luminescent Ln-MOF thin film for highly selective detection of nitroimidazoles in aqueous solutions based on inner filter effect. *J. Lumin.* **2019**, *205*, 23–29. [CrossRef]
25. Mao, X.; Liu, S.; Su, B.; Wang, D.; Huang, Z.; Li, J.; Zhang, Y. Luminescent europium(III)-organic framework for visual and on-site detection of hydrogen peroxide via a tablet computer. *Microchim. Acta* **2020**, *187*, 416. [CrossRef]
26. Yang, Y.; Zhao, L.; Sun, M.; Wei, P.; Li, G.; Li, Y. Highly sensitive luminescent detection toward polytypic antibiotics by a water-stable and white-light-emitting MOF-76 derivative. *Dye. Pigment.* **2020**, *180*, 108444. [CrossRef]
27. Guo, X.; Pan, Q.; Song, X.; Guo, Q.; Zhou, S.; Qiu, J.; Dong, G. Embedding carbon dots in Eu 3+ -doped metal-organic framework for label-free ratiometric fluorescence detection of Fe 3+ ions. *J. Am. Ceram. Soc.* **2020**, *104*, 886–895. [CrossRef]
28. Xiao, J.; Liu, M.; Tian, F.; Liu, Z. Stable Europium-based Metal–Organic Frameworks for Naked-eye Ultrasensitive Detecting Fluoroquinolones Antibiotics. *Inorg. Chem.* **2021**, *60*, 5282–5289. [CrossRef] [PubMed]
29. Yu, M.; Xie, Y.; Wang, X.; Li, Y.; Li, G. Highly Water-Stable Dye@Ln-MOFs for Sensitive and Selective Detection toward Antibiotics in Water. *ACS Appl. Mater. Interfaces* **2019**, *11*, 21201–21210. [CrossRef] [PubMed]
30. Zheng, L.; Zhu, T.; Xu, W.; Liu, L.; Zheng, J.; Gong, X.; Wudl, F. Solution-processed broadband polymer photodetectors with a spectral response of up to 2.5 μm by a low bandgap donor–acceptor conjugated copolymer. *J. Mater. Chem. C* **2018**, *6*, 3634–3641. [CrossRef]
31. Wang, B.; Lv, X.-L.; Feng, D.; Xie, L.-H.; Zhang, J.; Li, M.; Xie, Y.; Li, J.-R.; Zhou, H.-C. Highly Stable Zr(IV)-Based Metal–Organic Frameworks for the Detection and Removal of Antibiotics and Organic Explosives in Water. *J. Am. Chem. Soc.* **2016**, *138*, 6204–6216. [CrossRef]
32. Yang, M.; Tang, Q.; Meng, Y.; Liu, J.; Feng, T.; Zhao, X.; Zhu, S.; Yu, W.; Yang, B. Reversible "Off–On" Fluorescence of Zn2+-Passivated Carbon Dots: Mechanism and Potential for the Detection of EDTA and Zn2+. *Langmuir ACS J. Surf. Colloids* **2018**, *34*, 7767–7775. [CrossRef]
33. Miao, X.; Yan, X.; Qu, D.; Li, D.; Tao, F.F.; Sun, Z. Red Emissive Sulfur, Nitrogen Codoped Carbon Dots and Their Application in Ion Detection and Theraonostics. *ACS Appl. Mater. Interfaces* **2017**, *9*, 18549–18556. [CrossRef]
34. Lim, S.A.; Ahmed, M.U. A Simple DNA-based Electrochemical Biosensor for Highly Sensitive Detection of Ciprofloxacin Using Disposable Graphene. *Anal. Sci. Int. J. Jpn. Soc. Anal. Chem.* **2016**, *32*, 687–693. [CrossRef]
35. Xia, H.; Peng, M.; Li, N.; Liu, L. CdSe quantum dots-sensitized FRET system for ciprofloxacin detection. *Chem. Phys. Lett.* **2020**, *740*, 137085. [CrossRef]
36. Garrido, J.M.P.J.; Melle-Franco, M.; Strutyński, K.; Borges, F.; Brett, C.M.A.; Garrido, E.M.P.J. β–Cyclodextrin carbon nanotube-enhanced sensor for ciprofloxacin detection. *J. Environ. Sci. Health Part A* **2017**, *52*, 313–319. [CrossRef]
37. Attia, M.S.; Youssef, A.O.; Ismael, A.M.; Gaafer, R.; Adel, A.; Twfik, A.; Wafeey, A.; Afify, H.G.; Sayed, A. Highly Sensitive Eu^{3+} Doped in Sol-Gel Matrix Optical Sensor for The Assessment of Ciprofloxacin in Different Real Samples. *Egypt J. Chem.* **2018**, *61*, 121–129. [CrossRef]
38. Chang, C.-C.; Wang, G.; Takarada, T.; Maeda, M. Iodine-Mediated Etching of Triangular Gold Nanoplates for Colorimetric Sensing of Copper Ion and Aptasensing of Chloramphenicol. *ACS Appl. Mater. Interfaces* **2017**, *9*, 34518–34525. [CrossRef]
39. Li, Y.; Dai, H.; Feng, N.; Xie, X.; Zhang, J.; Li, W. Silver chloride nanoparticles-decorated molybdenum disulfide nanosheets for highly sensitive chloramphenicol detection. *Mater. Express* **2019**, *9*, 59–64. [CrossRef]

foods

MDPI

Article

A Smartphone Colorimetric Sensor Based on Pt@Au Nanozyme for Visual and Quantitative Detection of Omethoate

Biao Zhang [1,*], Ruofan Zhou [1], Huiqi Zhang [1], Danfeng Cai [1], Xiaodong Lin [2], Yihan Lang [1], Yulou Qiu [1], Xuping Shentu [1], Zihong Ye [1] and Xiaoping Yu [1]

[1] Zhejiang Provincial Key Laboratory of Biometrology and Inspection & Quarantine, College of Life Sciences, China Jiliang University, Xueyuan Street, Xiasha Higher Education District, Hangzhou 310018, China
[2] College of Food and Bioengineering, Zhengzhou University of Light Industry, Zhengzhou 450001, China
* Correspondence: zb@cjlu.edu.cn; Tel.: +86-136-2219-9129

Abstract: A smartphone colorimetric sensor based on the Pt@Au nanozyme was successfully developed for the visual and quantitative detection of omethoate in fruit and vegetables. The anti-omethoate antibody was conjugated on the surface of the Pt@Au nanozyme as a catalytic functional signal probe, and coating antigen conjugated on the surface of magnetic polystyrene microspheres (MPMs) was used as a separation capture probe. In the sensing system, when the catalytic functional signal probe was combined with a separation capture probe containing no omethoate, the visible blue color appeared with the addition of tetramethylbenzidine (TMB) chromogenic solution, and the maximum B value of the sensing system was obtained via the smartphone. With increasing concentrations of omethoate, the visualization of the sensing system decreased, and the B-value obtained via the smartphone dropped. Under optimal detection conditions, the omethoate could be detected in a linear range of 0.5–50 μg/L (R^2 = 0.9965), with a detection limit of 0.01 μg/L. The accuracy and reliability of the detection results of this colorimetric sensor were successfully confirmed by enzyme linked immunosorbent assay (ELISA) and gas chromatography. This colorimetric sensor provides a technical reference and potential strategy for the immunoassay of hazard factors in resource-scarce laboratories.

Keywords: colorimetric sensor; Pt@Au nanozyme; visualized biosensing; smartphone-based quantitative detection; omethoate

1. Introduction

Omethoate, as an organophosphorus pesticide, plays a very important role in improving the yield and quality of edible agricultural products [1]. Omethoate, as a highly efficient and toxic organophosphorus insecticide and acaricide, has mainly been used to effectively control various pests of cotton, wheat, fruit trees, vegetables, sorghum and other crops [2–5]. However, in order to pursue higher economic benefits, omethoate has been used unreasonably by some producers, and it can be sucked into the plant by the stems and leaves, and transmitted to various parts of the plant, which can easily result in omethoate residues [6–8]. Omethoate residue is difficult to degrade via ordinary microorganisms, and after ingestion by consumers, omethoate can inhibit the activity of acetylcholinesterase, affecting the function of human muscles and glands [9]. Meanwhile, omethoate also has serious impacts on food safety, the biological environment, and international trade [10–12]. According to GB 2763-2021 Maximum Residue Limits of Pesticides in Food, the allowable daily intake (ADI) of omethoate is 0.0003 mg/kg, the maximum allowable amount of residues in fruits, vegetables, cottonseed and wheat grains is 0.02 mg/kg, and the maximum allowable residue in soybean, sugar beet, sugarcane, tea, dry grain and multigrain cereals is 0.05 mg/kg [13]. Therefore, it is necessary to carry out strict testing, and control the usage and residual amounts of omethoate in accordance with regulations to reduce safety risks.

Many frequently used analytical approaches for omethoate have been reported, such as thin-layer chromatography (TLC) [14], high-performance liquid chromatography (HPLC) [15], gas chromatography (GC) [16], gas chromatography coupled with mass spectrometry (GC/MS) [17], and liquid chromatography coupled with mass spectrometry (LC-MS/MS) [18]. Although the above-mentioned mass spectrometry-related analytical methods are accurate, they have the disadvantages of being time consuming, high inspection and maintenance costs, and requiring professionals for operation [19]. Colorimetric sensors have been reported because of their high accuracy in the visual detection of target analytes without the use of complex equipment [20]. Many colorimetric sensors based on precious nanozymes have been prepared for the visual detection of organophosphorus pesticides, but the human eye is not sensitive to monochromatic or limited color changes, which also limits the application of colorimetric sensors in omethoate detection [21]. Therefore, it is necessary to develop a simple, visual, and quantitative colorimetric sensor based on a portable and inexpensive reading device for the detection of omethoate.

As natural biocatalysts, enzymes have high specificity and efficiency, and play important roles in biological processes such as signal transduction, metabolism, and digestion [22]. However, the production cost of natural enzymes is high, they are easy to denature and inactivate, and the yield is low, making them difficult to apply [23]. Therefore, researchers have gradually developed nanozymes with biomimetic enzyme capabilities and nanoscale dimensions [24]. The developed nanozymes have good stability, low production costs, and simple preparation and purification processes. At present, nearly a thousand nanomaterials with different compositions have been found to have enzyme-like activities. So far, a large number of nanomaterials, such as metal nanoparticles, metal oxide nanoparticles, carbon-based nanomaterials, and metal–organic framework materials, have been shown to have excellent catalytic properties for functioning like natural enzymes [25]. They have been widely used in cancer diagnosis and treatment, food safety detection, environmental monitoring, and other fields [26–28]. Therefore, compared with natural enzymes, the colorimetric sensor for omethoate based on the significant advantages of nanozymes are worth exploring.

In this work, the anti-omethoate antibody was conjugated on the surface of the catalytic functional Pt@Au nanoparticle to prepare a signal probe (Pt@Au@anti-omethoate antibody). A colorimetric sensor based on a signal probe and a smartphone was developed for the visual and quantitative detection of omethoate. Meanwhile, the specificity, practicability, and stability of the developed colorimetric sensor were evaluated. The colorimetric sensor is able to achieve the rapid, highly sensitive and low-cost detection of omethoate residue, providing a useful tool for the detection of omethoate residue in edible agricultural products, thus promoting food safety.

2. Materials and Methods

2.1. Reagents and Instruments

Omethoate, K_2PtCl_4, $HAuCl_4$, Pluronic F127, EDC, NHS and bovine serum albumin (BSA) were purchased from Sigma-Aldrich (Saint Louis, USA). Dimethoate, dichlorvos, methyl parathion, glyphosate, malathion, profenofos, acetamiprid, trichlorfon were obtained from Aladdin Biochemical Technology Co., Ltd. (Shanghai, China). Anti-omethoate monoclonal antibody was purchased from Kejie Industrial Development Co., Ltd. (Shenzhen, China). The Spotxel®Reader was obtained from the App Store (Sicasys Software GmbH, Germany).

2.2. Preparation of Pt@Au Nanozymes, Signal Probe and Capture Probe

Pt@Au nanoparticle nanozymes were synthesized using the liquid phase method with some modifications [29]. Firstly, 9.0 mL of 20 mM K_2PtCl_4 solution, 1.0 mL of 20 mM $HAuCl_4$ solution, and 104.0 mg of Pluronic F127 were added into a 25 mL round-bottomed flask, and the mixed solution was sonicated to dissolve completely. Then, 1.0 mL of 100 mM ascorbic acid (Vc) was added into the above solution. The mixed solution was sonicated in a water bath at 30 °C for 3 h. The mixed reaction solution was separated by centrifugation at 10,000 rpm/min for 10 min, and the precipitate was washed using ethanol and deionized water. Finally, Pt@Au nanozyme precipitate was reconstituted with 0.9 mL of deionized water and stored at 4 °C until use.

The Pt@Au@Antibody was prepared as follows. Briefly, 50 μg of anti-omethoate antibody was added to 3.0 mL of 15-fold-diluted Pt@Au solution with pH 9.0, and the mixed solution was incubated at 4 °C for 4 h. Then, 30 μL of PBS (0.01 M, pH 7.4, containing 10% BSA) was added to block the excess sites on the surface of Pt@Au. The solution was centrifuged at 10,000 rpm/min for 20 min to remove unconjugated anti-omethoate antibody and BSA. Finally, the Pt@Au@Antibody of the catalytic signal probe was obtained by reconstitution in 1.0 mL of PBS (0.01 M, pH 7.4, 10% BSA) and stored at 4 °C for later use.

The capture probe was prepared as follows. Briefly, 10.0 mg EDC and 5.0 mg NHS were added into 1.0 mL of 5.0 mg/mL magnetic polystyrene microspheres (MPMs) to activate the carboxyl groups on the surface. Then, 50 μg of omethoate coating antigen (OCA) was added into the solution to incubate for 1 h, and 30 μL phosphate buffer saline (PBS) (0.01 M, pH 7.4, containing 10% BSA) was used to block excess MPM epitopes. The supernatant was discarded by external magnetic force, and the conjugation of coating antigen and MPMs (OCA@MPMs) was washed. Finally, the capture probe was prepared by reconstituting the precipitate into 5.0 mL PBS (0.01 M, pH 7.4).

2.3. Development of Colorimetric Sensor

The construction process of the colorimetric sensor based on a smartphone was as follows. For the detection system, 50 μL of the capture probe and 50 μL of omethoate standard solution (0, 0.05, 0.1, 0.5, 1.0, 5.0, 10, 50, 100, 200, 500 μg/L) were mixed in a 1.5 mL tube. Then, 50 μL of the catalytic signal probe was added, and the solution was incubated for 20 min at room temperature. The supernatant was discarded by magnetic separation, and the immunoprecipitated complex (Pt@Au@Antibody-OCA@MPMs) was washed with PBS (0.01 M, pH 7.4) in the tube. In addition, 400 μL of TMB chromogenic solution was added into the above tube, and the color reaction time was 5 min at room temperature. The B value of the detection system was captured using Spotxel®Reader on the smartphone, and a linear equation was constructed according to the relationship between the B value and the concentration of the omethoate standard solution. Finally, the specificity and applicability of the colorimetric sensor was evaluated by using structural analogs and similar substances, and actual samples testing, respectively.

2.4. Sample Preparation

The samples were pretreated for the developed colorimetric sensor and ELISA according to the national standard GB/T 23200.116-2019 with some modification. Briefly, 20.00 g of Chinese cabbage and small Chinese cabbage samples were put into 150 mL beakers with 40 mL of acetonitrile respectively. The samples were homogenized at high speed for 3 min and the mixture was filtered into a 100 mL centrifuge tube with 6.50 g sodium chloride. The mixed sample solution was vigorously shaken for 2 min and let stand for 30 min. Then, 10 mL of the supernatant was pipetted into a colorimetric tube and dried with nitrogen. One milliliter of pH = 7.0 PBS buffer solution containing 5% methanol was used to dissolve the residue, and the solution was filtered with a microporous membrane (0.22 × 25 mm) for later analysis. For GC analysis, the other steps are basically the same as the above, except that 5.0 mL of acetone was used to reconstitute the sample residue after nitrogen blowing, and then the 25-fold diluted sample solution was measured.

2.5. ELISA and GC Analysis

The detailed procedure of ELISA for omethoate was listed in Supplementary Material. The omethoate concentration in samples was measured by GC 2010 Plus equipped with flame photometric detector. Then, 50% of polyphenylmethylsiloxane quartz capillary column (30 m × 0.53 mm (inner diameter) × 1.0 μm) was employed for the detection of omethoate. The applied parameters for the detection conditions were: column temperature, 150 °C for 2 min, 8 °C/min programmed temperature increase to 210 °C, 5 °C/min programmed temperature increase to 250 °C, maintained for 15 min, with nitrogen as the carrier gas at a flow rate of 8.4 mL/min. The temperatures of the injection port and the detector were 250 °C and 300 °C, respectively. The injection volume was 1.0 μL.

3. Results and Discussion

3.1. Principle of the Colorimetric Sensor

The preparation of the signal probe, capture probe and detection process is shown in Scheme 1. K_2PtCl_4, $HAuCl_4$, Pluronic F127 and Vc were employed to prepare the Pt@Au nanozymes, and MPMs and coating antigen of omethoate were used to prepare the capture probe (Scheme 1A). The capture probe, the omethoate standard (or sample) solution, and the signal probe are dropped into a 1.5 mL centrifuge tube and incubated for 20 min. As shown in Scheme 1B, the supernatant containing the Pt@Au@antibody-omethoate (or Pt@Au@antibody) was discarded under external magnetic force. Then, TMB solution was added into the tube containing the immunoprecipitated complex (Pt@Au@Antibody-OCA@MPMs). In the absence of omethoate molecules, the color visible to the naked eye was the bluest, and the B value based on the smartphone reading was the highest. With the increase of omethoate concentration in the detection system, the less signal probes captured by the capture probe, the visible color by the naked eye was lighter, and the B value obtained by the smartphone was relatively lower. According to the obtained B value and the concentrations of omethoate, the quantitative detection standard curve for omethoate was constructed.

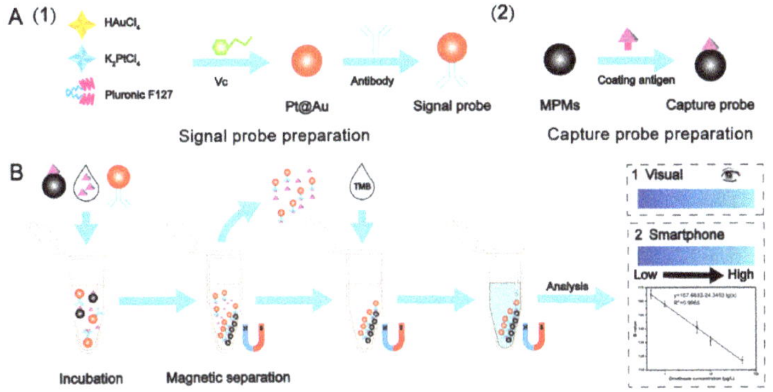

Scheme 1. Schematic illustration of preparation of signal probe and capture probe (**A**), and the detection process of proposed colorimetric sensor (**B**).

3.2. Characterization of Pt@Au Nanozyme

In this work, Pt@Au nanoparticles with catalytic function were prepared using the liquid phase method. As shown in Figure 1A, the Pt@Au nanoparticles are spherical in Pt@Au TEM image, and the average particle size of Pt@Au Pt@Au nanoparticles is 19.0 ± 5.6 nm, on the basis of statistical analysis. The distribution of Au and Pt elements was analyzed, and elemental mapping images (Figure 1B–D) show the distribution of Au and Pt elements in Pt@Au nanoparticles. The uniform distribution of Pt and Au elements shows that Pt@Au nanoparticles have both the catalytic function of Pt and the ability of Au coupled to antibodies. The XPS spectra of Pt 4f and Au 4f are shown in Figure 1E,F, the binding energy peaks at 70.73 eV and 74.08 eV correspond to Pt0 $4f_{7/2}$ and Pt0 $4f_{5/2}$, and the peak at 86.81 eV reflects Au0 $4f_{5/2}$. The results are basically consistent with those reported in the literature [30], which indicates that the Pt@Au nanoparticles had been successfully prepared. Therefore, Pt@Au nanoparticles can be applied to prepare the catalytic signal probe, based on which the colorimetric sensor for omethoate was constructed.

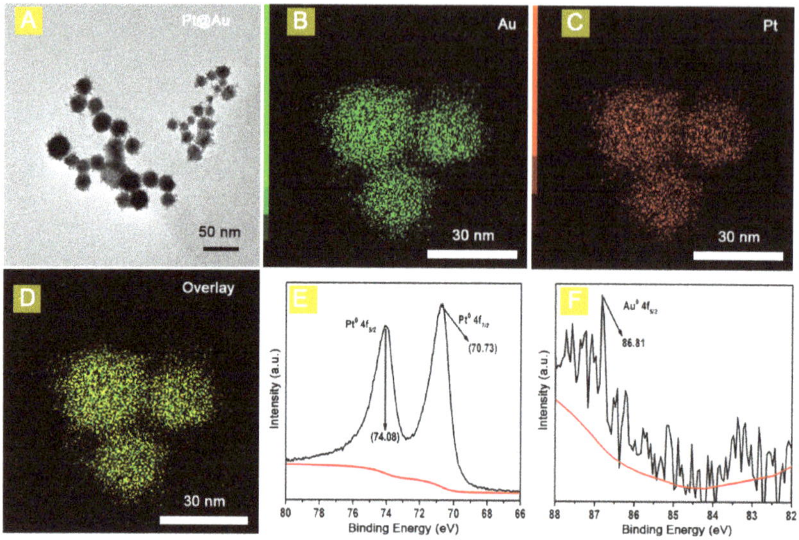

Figure 1. TEM images of Pt@Au (**A**); elemental mapping images of Au (**B**), Pt (**C**), and Au and Pt (**D**) in Pt@Au nanoparticles; and high-resolution XPS spectra of Pt4f (**E**) and Au4f (**F**).

3.3. Development of Colorimetric Sensor

To obtain better detection performance with the colorimetric sensor based on a smartphone, the amount of anti-omethoate antibody added in the preparation of signal probes, the amount of coating antigen added in the preparation of the capture probes, the amount of signal probe added, and the incubation time were optimized. The amounts of added anti-omethoate antibody and coating antigen were optimized by the conjugation amount using the BCA protein quantification kit. The signal probe and incubation time were optimized based on the B value. The results are shown in Figure 2, and the optimal dosages of anti-omethoate antibody and coated antigen were 50 μg and 50 μg, respectively, and the corresponding conjugation ratios were 52.80% and 76.60%. Under the above conditions, two probes were prepared for the colorimetric sensor. According to the B value of the detection system, the optimal added amount of Pt@Au@Antibody, and the incubation time for the development of the colorimetric sensor are 50 μL and 15 min, respectively.

Under the above experimental conditions, the omethoate standard solution with different concentrations (0, 0.05, 0.1, 0.5, 1.0, 5.0, 10, 50, 100, 200, 500 μg/L) was detected by the developed colorimetric sensor. The immunoprecipitated complex (Pt@Au@Antibody-OCA@MPMs) was separated by external magnetic force and 400 μL of TMB mixture solution was added to the above complex system. After color reaction for 5 min, the B value of the detection system was obtained in a dark box using the SpotxelRReader software. The dark box device was prepared by transforming an LED photo studio (Figure S1). There is a circle of lights on the top of the dark box device, and the lights can be lit by a mobile power supply or a socket power supply. As shown in Figure 3, the B value shows a decreasing trend with increasing omethoate concentration (0.5, 1.0, 5.0, 10, 50 μg/L), which is due to the competition between omethoate molecules and the capture probe with respect to binding to the catalytic signal probe. The visual detection results regarding the presence of different omethoate concentrations (0, 0.05, 0.1, 0.5, 1.0, 5.0, 10, 50, 100, 200, 500 μg/L) are shown in Figure S2 in the Supplementary Materials. The linear detection equation for omethoate is y = 157.6633-24.3453 lg(x) (R^2 = 0.9965), and the linear detection range is 0.5 to 50 μg/mL, with a detection limit of 0.01 μg/mL. To evaluate the sensitivity of the developed colorimetric sensor under the working conditions (Table S1), the same anti-omethoate antibody and coating antigen were employed to establish traditional indirect competitive ELISA for the detection of omethoate. The standard inhibition curve of the ELISA for omethoate is shown in Figure S3. The limits of detection (IC_{15}) and sensitivity (IC_{50}) of ELISA for omethoate are 5.54 μg/L and 131.17 μg/L, respectively. Compared with the ELISA for omethoate, the sensitivity of the developed colorimetric sensor is improved by two orders of magnitude. Compared with the other methods for detecting omethoate presented in the literature [31–34], this proposed colorimetric sensor improves the specificity and detection speed for omethoate (Table S2).

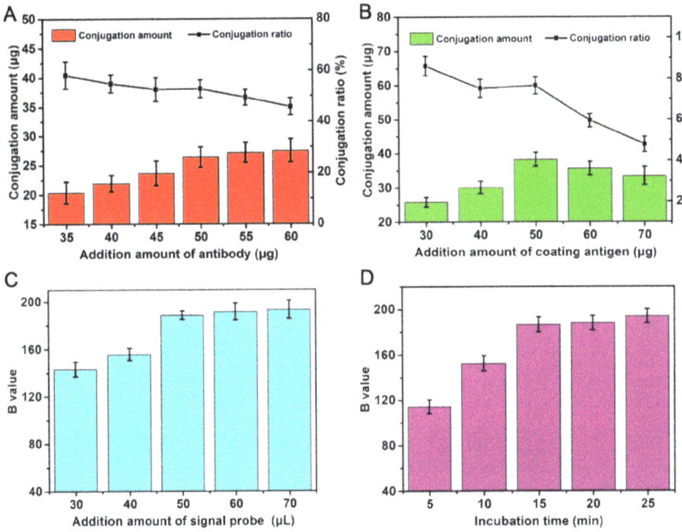

Figure 2. The optimization of the working parameters of the colorimetric sensor. (**A**) Added amount of anti-omethoate antibody, (**B**) added amount of coating antigen, (**C**) added amount of signal probe, and (**D**) incubation time.

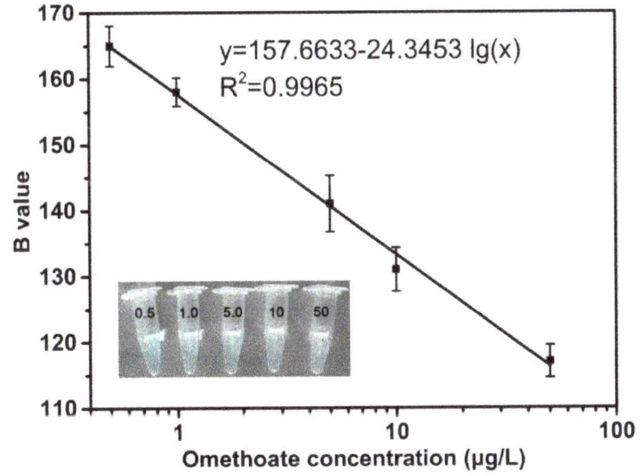

Figure 3. The standard curve of the developed colorimetric sensor for omethoate; and inset: color development images at omethoate concentrations of 0.5~50 μg/L.

3.4. Evaluation of Specificity

The evaluation of specificity is an important indicator for an immunoassay. In this work, dimethoate, dichlorvos, methyl parathion, glyphosate, malathion, profenofos, acetamiprid, and trichlorfon, as functional and structural analogs of omethoate, were employed to evaluate the specificity of the colorimetric sensor. As shown in Figure 4, the ΔB was 15 and 75 when the concentrations of omethoate were 0.1 μg/L and 10 μg/L, and the difference in the B value (ΔB) between the sample without omethoate and 1000 μg/L of different analogs of omethoate, respectively. The inset is the structure diagram of the corresponding structural analog of omethoate. Meanwhile, the ΔB values for dimethoate, dichlorvos, methyl parathion, glyphosate, malathion, profenofos, acetamiprid, and trichlorfon were less than 10. These results indicate that the anti-omethoate antibody specifically recognized omethoate, and that the constructed colorimetric sensor, with a good specific recognition for omethoate, would not be affected by the functional and structural analogs of omethoate. The constructed colorimetric sensor is able to analyze omethoate with high specificity in real samples.

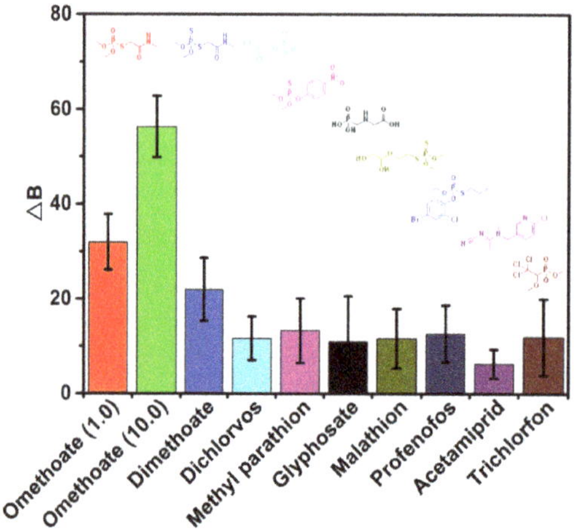

Figure 4. Specificity analysis of the developed colorimetric sensor.

3.5. Analytical Application in Food Samples

To evaluate the practicability and stability of the developed colorimetric sensor, we selected Chinese cabbage and tomato to measure the background concentration and the different concentration levels of the added omethoate in actual samples. The samples were spiked with three concentration levels of omethoate: 20 μg/kg, 200 μg/kg, and 1000 μg/kg. The results obtained when detecting omethoate in Chinese cabbage and tomato using the developed colorimetric sensor were confirmed by GC and ELISA based on the same antibody and coating antigen of omethoate. As shown in Table 1, the concentrations of omethoate in the non-spiked Chinese cabbage and tomato were not detected using the developed method, ELISA, or GC. With the addition of omethoate into Chinese cabbage at concentrations of 20 μg/kg, 200 μg/kg, and 1000 μg/kg, the results detected by the developed assay were 17.67 μg/kg, 178.83 μg/kg and 902.00 μg/kg, respectively, and the results detected by ELISA (or GC) were 18.40 μg/kg, 177.97 μg/kg, and 895.60 μg/kg (or 19.03 μg/kg, 189.50 μg/kg, and 905.20 μg/kg). The same level gradient concentrations of omethoate were added into the tomato, and the detection results using the developed assay were 17.70 μg/kg, 172.30 μg/kg and 803.47 μg/kg, and the detection results using ELISA (or GC) were 17.53 μg/kg, 168.83 μg/kg, and 854.80 μg/kg (or 17.51 μg/kg, 177.83 μg/kg, and 897.79 μg/kg), respectively. The above results show that the results detected by the developed colorimetric sensor are in good agreement with those detected by GC and ELISA (R > 0.99). The stability of the results detected using the developed colorimetric sensor was evaluated using the coefficient of variation (CV). The CV value of the colorimetric sensor (5.16–11.47%) is slightly larger than that of ELISA (4.10–6.19%) and GC (2.12–4.33%), indicating that the accepted stability of the colorimetric sensor is not as good as that of ELISA and GC, and subsequent work can focus on improving the stability of the colorimetric sensor. Meanwhile, the colorimetric sensor could also be a selective tool for the detection of omethoate in fruits and vegetables.

Table 1. Detection results of omethoate in Chinese cabbage and tomato (*n* = 10).

Samples	Added (μg/kg)	This Colorimetric Sensor			ELISA			GC		
		Detected (μg/kg)	Recovery (%)	CV (%)	Detected (μg/kg)	Recovery (%)	CV (%)	Detected (μg/kg)	Recovery (%)	CV (%)
Chinese cabbage	0	ND	-	-	ND	-	-	ND	-	-
	20	17.67	88.33	10.92	18.40	92.00	4.10	19.03	95.17	2.12
	200	178.83	89.42	9.61	177.97	88.98	6.02	189.50	94.75	3.71
	1000	902.00	90.20	5.16	895.60	89.56	6.19	905.20	90.52	4.31
Tomato	0	ND	-	-	ND	-	-	ND	-	-
	20	17.70	88.50	11.47	17.53	87.67	5.71	17.51	87.57	4.33
	200	172.30	86.15	8.13	168.83	84.42	5.78	177.83	88.92	3.37
	1000	803.47	80.35	6.02	854.80	85.48	6.15	897.79	89.78	3.23

CV, coefficient of variation; ND, not detected; -, not be calculated.

4. Discussion

In this work, the anti-omethoate antibody conjugated to the surface of the prepared Pt@Au nanozyme as a catalytic signal probe was used to construct a colorimetric sensor. The developed colorimetric sensor based on the smartphone was able to accurately and rapidly detect omethoate in Chinese cabbage and tomato. Meanwhile, the constructed colorimetric sensor could be used to perform semi-quantitative visual detection of omethoate by means of the naked eye, without sophisticated detection instruments. The constructed colorimetric sensor has a linear detection of omethoate in the range of 0.5–50 μg/L, with a detection limit of 0.01 μg/L. The accuracy and reliability of this colorimetric sensor was successfully confirmed by ELISA and GC, and the detection results obtained by this colorimetric sensor are in good agreement with the results detected using ELISA and GC. In conclusion, a convenient, intelligent, and highly sensitive detection method for omethoate is presented in this work, demonstrating that a strategy based on smartphone and Pt@Au nanozyme could be successfully applied to the detection and monitoring of pesticide residues in other fruits and vegetables.

Supplementary Materials: The following supporting information can be downloaded at: https://www.mdpi.com/article/10.3390/foods11182900/s1, Figure S1: the dark box device diagrams in different locations; Figure S2: visual results for the different concentrations of omethoate; Figure S3: standard inhibition curve of the ELISA for omethoate in PBS; Table S1: optimal working conditions of ELISA; Table S2: Comparison of sensors for omethoate detection.

Author Contributions: Conceptualization, methodology, writing—original draft preparation, funding acquisition, B.Z.; data curation, R.Z., H.Z. and D.C.; validation, Y.L. and Y.Q.; conceptualization, supervision, X.L.; project administration, writing—review and editing, funding acquisition, X.S., Z.Y. and X.Y. All authors have read and agreed to the published version of the manuscript.

Funding: This work was supported by the Talents Scientific Research Startup Project of China Jiliang University (01101-210952), the Research award funds of the College of Life Sciences, China Jiliang University (grant number 00106-55), the Key R&D Program of Zhejiang Province (2021C02061, 2022C02049), and Leading Talents in Science and Technology Innovation of Ten Thousand Talents Program in Zhejiang Province (2019R52018).

Data Availability Statement: The datasets generated for this study are available on request from the corresponding author.

Conflicts of Interest: The authors declare that they have no known competing financial interests or personal relationships that could have appeared to influence the work reported in this paper.

References

1. Zhang, C.P.; Lin, B.X.; Cao, Y.J.; Guo, M.L.; Yu, Y. Fluorescence Determination of Omethoate Based on a Dual Strategy for Improving Sensitivity. *J. Agric. Food Chem.* **2017**, *65*, 3065–3073. [CrossRef] [PubMed]
2. Chen, J.; Dong, D.; Ye, S. Detection of pesticide residue distribution on fruit surfaces using surface-enhanced Raman spectroscopy imaging. *RSC Adv.* **2018**, *8*, 4726–4730. [CrossRef] [PubMed]
3. Huang, S.Y.; Guo, M.L.; Tan, J.A.; Geng, Y.Y.; Wu, J.Y.; Tang, Y.W.; Su, C.C.; Lin, C.C.; Liang, Y. Novel Fluorescence Sensor Based on All-Inorganic Perovskite Quantum Dots Coated with Molecularly Imprinted Polymers for Highly Selective and Sensitive Detection of Omethoate. *ACS Appl. Mater. Interfaces* **2018**, *10*, 39056–39063. [CrossRef] [PubMed]
4. Herron, G.A.; Wilson, L.J. Can resistance management strategies recover insecticide susceptibility in pests?: A case study with cotton aphid *Aphis gossypii* (Aphididae: Hemiptera) in Australian cotton. *Austral Entomol.* **2017**, *56*, 1–13. [CrossRef]
5. Zhang, L.; Rao, Z.; Ji, H. NIR Hyperspectral Imaging Technology Combined with Multivariate Methods to Study the Residues of Different Concentrations of Omethoate on Wheat Grain Surface. *Sensors* **2019**, *19*, 3147. [CrossRef] [PubMed]
6. Tao, Y.G.; Wang, Y.M.; Yan, S.L.; Ye, L.B. Optimization of omethoate degradation conditions and a kinetics model. *Int. Biodeter. Biodegr.* **2008**, *62*, 239–243.
7. Gavahian, M.; Khaneghah, A.M. Cold plasma as a tool for the elimination of food contaminants: Recent advances and future trends. *Crit. Rev. Food Sci.* **2020**, *60*, 1581–1592. [CrossRef] [PubMed]
8. Yu, L.L.; Wang, L.N.; Zhao, Y.; Wang, B.J. Identification and Dissipation of Omethoate and Its Main Metabolite DMP in Wheat Determined by UPLC-QTOF/MS. *J. Agric. Food Chem.* **2019**, *67*, 5891–5898. [CrossRef]
9. Sidhu, G.K.; Singh, S.; Kumar, V.; Dhanjal, D.S.; Datta, S.; Singh, J. Toxicity, monitoring and biodegradation of organophosphate pesticides: A review. *Crit. Rev. Env. Sci. Tec.* **2019**, *49*, 1135–1187. [CrossRef]
10. Lazarevi Pašti, T.D.; Pašti, I.A.; Joki, B.; Babića, B.M.; Vasić, V.M. Heteroatom-doped mesoporous carbons as efficient adsorbents for removal of dimethoate and omethoate from water. *RSC Adv.* **2016**, *6*, 62128–62139. [CrossRef]
11. Nair, R.V.; Chandran, P.R.; Mohamed, A.P.; Pillai, S. Sulphur-doped graphene quantum dot based fluorescent turn-on aptasensor for selective and ultrasensitive detection of omethoate. *Anal. Chim. Acta* **2021**, *1181*, 338893. [CrossRef] [PubMed]
12. Bouagga, A.; Chaabane, H.; Toumi, K.; Hamdane, A.M.; Nasraoui, B.; Joly, L. Pesticide residues in Tunisian table grapes and associated risk for consumer's health. *Food Addit. Contam. B* **2019**, *12*, 135–144. [CrossRef] [PubMed]
13. *GB/T 2763-2021*; National Food Safety Standard Maximum Residue Limits of Pesticides in Foods (Print Version). China Agricultural Press: Beijing, China, 2021. Available online: http://down.foodmate.net/standard/sort/3/97819.html (accessed on 1 September 2022).
14. Huo, Q.; Quan, Q.; Tao, F.; Wang, Y. Controlling the Red Spider Mite, Tetranychus urticae Koch with Ailanthus altissima Leaf Extract. *Egypt. J. Biol. Pest Control* **2016**, *26*, 671–674.
15. Zheng, L.L.; Pi, F.W.; Wang, Y.F.; Xu, H.; Zhang, Y.Z.; Sun, X.L. Photocatalytic degradation of Acephate, Omethoate, and Methyl parathion by Fe_3O_4@SiO_2@mTiO_2 nanomicrospheres. *J. Hazard. Mater.* **2016**, *315*, 11–22. [CrossRef]
16. Yu, R.; Wang, Y.; Cui, Z.W.; Xu, G.H.; Guan, Z.Y.; Yu, Y.; Liu, J.S. Human health risk assessment of organophosphorus pesticides in maize (*Zea mays* L.) from Yushu, Northeast China. *Hum. Ecol. Risk Assess.* **2018**, *24*, 642–652. [CrossRef]
17. Osaili, T.M.; Al Sallagi, M.S.; Dhanasekaran, D.K.; Odeh, W.A.M.B.; Al Ali, H.J.; Al Ali, A.A.S.A.; Radwan, H.; Obaid, R.S.; Holley, R. Pesticide residues in fresh vegetables imported into the United Arab Emirates. *Food Control* **2021**, *133*, 108663. [CrossRef]
18. Pasar, R.; Pallavi, M.S.; Naik, R.H.; Devaraj, M.; Nandini, P.; Bheemanna, M.; Badariprasad, P.R.; Paramasivam, M. Simultaneous determination of dimethoate and its metabolite omethoate in curry leaf using LC-MS/MS and risk assessment. *J. Sep. Sci.* **2022**, *45*, 1831–1838. [CrossRef]
19. Tian, F.Y.; Zhou, J.; Jiao, B.N.; He, Y. A nanozyme-based cascade colorimetric aptasensor for amplified detection of ochratoxin A. *Nanoscale* **2019**, *11*, 9547–9555. [CrossRef]
20. Zhang, Q.T.; Li, M.F.; Guo, C.Y.; Jia, Z.; Wan, G.C.; Wang, S.F.; Min, D.Y. Fe_3O_4Nanoparticles Loaded on Lignin Nanoparticles Applied as a Peroxidase Mimic for the Sensitively Colorimetric Detection of H_2O_2. *Nanomaterials* **2019**, *9*, 210. [CrossRef]
21. Zhang, Q.T.; Yu, Y.Y.; Yun, X.J.; Luo, B.; Jiang, H.R.; Chen, C.Z.; Wang, S.F.; Min, D.Y. Multicolor Colorimetric Sensor for Detection of Omethoate Based on the Inhibition of the Enzyme-Induced Metallization of Gold Nanorods. *ACS Appl. Nano Mater.* **2020**, *3*, 5212–5219. [CrossRef]
22. Zhang, X.D.; Chen, X.K.; Zhao, Y.L. Nanozymes: Versatile Platforms for Cancer Diagnosis and Therapy. *Nano-Micro Lett.* **2022**, *14*, 95. [CrossRef] [PubMed]
23. Wu, J.J.X.; Wang, X.Y.; Wang, Q.; Lou, Z.P.; Li, S.R.; Zhu, Y.Y.; Li, Q.; Wei, H. Nanomaterials with enzyme-like characteristics (nanozymes): Nextgeneration artificial enzymes (II). *Chem. Soc. Rev.* **2019**, *48*, 1004–1076. [CrossRef] [PubMed]
24. Wang, X.; Hu, Y.; Wei, H. Nanozymes in bionanotechnology: From sensing to therapeutics and beyond. *Inorg. Chem. Front.* **2016**, *3*, 41–60. [CrossRef]
25. Zhong, H.T.; Xue, Y.T.; Zhang, P.Y.; Liu, B.; Zhang, X.; Chen, Z.B.; Li, K.; Zheng, L.R.; Zuo, X. Cascade reaction system integrating nanozymes for colorimetric discrimination of organophosphorus pesticides. *Sens. Actuators B-Chem.* **2022**, *350*, 130810. [CrossRef]
26. Huang, Y.; Ren, J.; Qu, X. Nanozymes: Classification, catalytic mechanisms, activity regulation, and applications. *Chem. Rev.* **2019**, *119*, 4357–4412. [CrossRef] [PubMed]
27. Liang, M.; Yan, X. Nanozymes: From new concepts, mechanisms, and standards to applications. *Acc. Chem. Res.* **2019**, *52*, 2190–2200. [CrossRef]

28. Gomaa, E.Z. Nanozymes: A Promising Horizon for Medical and Environmental Applications. *J. Clust. Sci.* **2022**, *33*, 1275–1297. [CrossRef]
29. Ataee-Esfahani, H.; Imura, M.; Yamauchi, Y. All-Metal Mesoporous Nanocolloids: Solution-Phase Synthesis of Core-Shell Pd@Pt Nanoparticles with a Designed Concave Surface. *Angew. Chem. Int. Ed.* **2013**, *52*, 13611–13615. [CrossRef]
30. Torres-Pacheco, L.J.; De Leon-Rodriguez, A.; Alvarez-Contreras, L.; Guerra-Balcazar, M.; Arjona, N. Sorbitol electro-oxidation reaction on sub <10 nm PtAu bimetallic Nanoparticles. *Electrochim. Acta* **2020**, *353*, 136593.
31. Oujji, N.B.; Bakas, I.; Istamboulie, G.; Ait-Ichou, I.; Ait-Addi, E.; Rouillon, R.; Noguer, T. A simple colorimetric enzymatic-assay, based on immobilization of acetylcholinesterase by adsorption, for sensitive detection of organophosphorus insecticides in olive oil. *Food Control* **2014**, *46*, 75–80.
32. Han, Z.; Chi, C.S.; Bai, B.; Liu, G.; Rao, Q.X.; Peng, S.J.; Liu, H.; Zhao, Z.H.; Zhang, D.B.; Wu, A.B. Chromogenic platform based on recombinant drosophila melanogaster acetylcholinesterase for visible unidirectional assay of organophosphate and carbamate insecticide residues. *Anal. Chim. Acta* **2012**, *720*, 126–133.
33. Guo, X.S.; Zhang, X.Y.; Cai, Q.; Shen, T.; Zhu, S.M. Developing a novel sensitive visual screening card for rapid detection of pesticide residues in food. *Food Control* **2013**, *30*, 15–23.
34. Pang, S.; Labuza, T.P.; He, L. Development of a single aptamer-based surface enhanced raman scattering method for rapid detection of multiple pesticides. *Analyst* **2014**, *139*, 1895–1901

foods

MDPI

Article

Facile Synthesis of N, S-Doped Carbon Quantum Dots from Food Waste as Fluorescent Probe for Sensitive Detection of Thiamphenicol and Its Analogues in Real Food Samples along with an Application in Bioimaging

Shujuan Chen [1,*], Wanlin Ouyang [1,2], Yiting Zhu [1], Li He [1], Likou Zou [3], Xiaolin Ao [1], Shuliang Liu [1], Yong Yang [1] and Jianlong Li [1]

[1] College of Food Science, Sichuan Agricultural University, Ya'an 625014, China
[2] Yi'yang Agricultural Products Quality Inspection and Testing Center, Yi'yang 413000, China
[3] College of Resources, Sichuan Agricultural University, Chengdu 611130, China
* Correspondence: chenshujuan1@163.com; Tel.: +86-0835-2882311

Abstract: Herein, N, S co-doped carbon quantum dots (N, S-CDs) with high absolute quantitative yield (Abs-QY) of 50.2% were produced by hydrothermal treatment of food residue crayfish shells. A new detection method of thiamphenicol (TAP) and its analogues was established by discovering the obvious fluorescence response between TAP and N, S-CDs, which achieved a wide linear range of 20–300 $\mu g \cdot L^{-1}$ with a detection limit (LOD) of 11.12 $\mu g \cdot L^{-1}$. This novel probe exhibited strong sensitivity and shows rapid response in complex food matrices (overall detection time is less than 45 min) mainly induced by static quenching. Spiked food sample recovery ranged from 97.3 to 99.34%. Further, the cell experiments of N, S-CDs were conducted, and the cell viability remained 91.76% under high concentration of N, S-CDs due to the environmentally friendly materials. The low cytotoxicity and good cytocompatibility make these N, S-CDs compatible for cell bioimaging and intracellular detection of TAP.

Keywords: food waste; N, S-CDs; rapid detection; thiamphenicol; bioimaging

1. Introduction

Amphenicol antibiotics are a kind of broad-spectrum antibiotic drugs [1], used widely for the veterinary treatment of food-producing animals. Chloramphenicol (CAP) represents amphenicol antibiotics, but it was prohibited in the European Community Regulations 1430/94 notice in 1994 [2]. Thiamphenicol (TAP) is similar in structure to CAP but much more potent [3] and is a perfect substitute for CAP in treating diseases of poultry and animals [4]. Several studies have documented that TAP has a broad spectrum of action against bacterial infections of the gastrointestinal and respiratory tract, bacterial prostatitis, and sexually transmitted diseases. Given the excellent antibacterial properties and fungibility, TAP is an attractive option in many countries for treating animal diseases. However, the overuse of TAP in animal feedstuffs can cause the presence of its residues in animal-derived foods and induce antibacterial resistance [5]. Moreover, it shows hematological toxicity, which presents a danger to human health. Hence, the European Economic Community (EEC) [6] has defined the maximum residue limit (MRL) of TAP in milk and poultry muscle tissues as 50 $\mu g \cdot kg^{-1}$. Currently, the conventional detection methods for TAP involve enzyme-linked immunosorbent (ELISA) [7] assay, UV spectrophotometry [8], high-performance liquid chromatography (HPLC) [9], and high-performance liquid chromatography–mass spectrometry (LC-MS) [10]. Among these, ELISA can rapidly achieve trace detection and cope with large number of samples but is still prone to false positives or negatives. The UV spectrophotometry is simple in operation but has a poor detection limit and selectivity. The HPLC and LC-MS method has its advantages of good sensitivity and accuracy; nevertheless, it requires long analysis time, tedious pretreatment steps, and many organic solvents. Considering that, it is vital to establish a novel method for the highly effective determination of TAP residues in food samples.

Fluorescence sensing detection method has been widely used due to excellent optical performance and high sensitivity in recent years. Several studies have shown that TAP detection techniques by semiconductor metal-quantum dot can obtain acceptable linear range and detection [11,12]. However, semiconductor quantum dots are primarily composed of II–VI or III–V metal elements, which still have biosecurity threats. Surprisingly, carbon quantum dots (CDs) display excellent photostability, favorable biocompatibility, and low toxicity, which is an alternative to the traditional semiconductor-quantum dots gradually [13]. Up to now, many natural food products are easily available to replace the usual chemicals for preparing CDs. The efficient utilization of food waste with rich carbon sources is a more favorable alternative option, with advantages of being cost-effective and eco-friendly [13,14]. CDs synthesized from natural resources still pose significant challenges because of their low quantum yield and poor stability. One of the most effective strategies to regulate the photoluminescence properties and quantum yield of CDs is heteroatom doping. For instance, nitrogen- and sulfur-doped CDs (N, S-CDs) possess abundant nitrogen- and sulfur-containing surface functional groups, which may bring better features and broader application prospects for the fluorescent materials [15]. Thus far, there have been several studies on the synthesis of N, S-CDs for applications such as detection of metal ion in lake water and environmental water (Table S1) [16–22]. However, the N, S-CDs prepared in these studies required special analytical grade reagents, and the fluorescence quantum yield and cytotoxicity were not satisfactory. In addition, these researches lacked research on the stability of probes as well as more detailed discussion of the fluorescence quenching mechanism. Dong prepared CDs, N-CDs, and N, S-CDs with different dopants, and among them, N, S-CDs showed better fluorescence (FL), quantum yield (QY), and fluorescence lifetime [23], which provides us with an important scientific reference for the FL enhancement mechanism of CDs (such as the choice of effective dopants). However, the carbon precursor used by Dong is the universally applicable citric acid, and we believe that replacing this reagent with food waste is more economical and environmentally friendly. Given the low weight-gain risk and delicious taste of crayfish, the demand for crayfish is increasing. According to the survey, "crayfish meat accounts for only 30% of total crayfish weight, while about 420,000~434,000 tons of crayfish shell are produced per year as waste" [24], and it causes serious environmental problems. Crayfish shell is rich in protein and chitin, which meets the carbon source requirement of CDs and inspires us to fully develop and utilize crayfish shell waste in an innovative way. According to report, food waste has previously been introduced as a carbon precursor in carbon dots synthesis [13]. Therefore, in order to verify the unique advantages of fluorescent probes prepared from crayfish shells for detecting TAP, we conducted a large number of preliminary experiments to screen food waste. As shown in Table S2, firstly, compared with other food wastes, the fluorescent probes prepared from crayfish shells have better FL intensities and exhibit better fluorescence responses to TAP. Secondly, compared to the probe without dopant, the probe with dopant, especially L-cysteine, showed the best detection effect for TAP. However, as far as we know, there are no reports for preparing N, S-CDs from crayfish shell waste; notably, this environmental-friendly probe has not been practically used to detect TAP in real food and living cells. It is of great significance for turning waste into value-added products and improving the comprehensiveness of TAP testing methods.

Herein, a new method of synthesis based on N, S-CDs is proposed. The readily available crayfish shell waste and L-cysteine were employed as a precursor and dopant for the hydrothermal synthesis of CDs, respectively. The synthesis mechanism diagram is shown in Figure 1. The novel fluorescent materials display a better Abs-QY and stable fluorescent lifetime, and thus, the probe can be effectively quenched by TAP via static quenching. The results demonstrate that the method our proposed can rapidly detect trace TAP in food samples within 45 min without interference from complex environment (including sample pretreatment time). Importantly, the cell viability remained 91.76% at the high concentration of N, S-CDs, and TAP can be also smoothly tracked by N, S-CDs in living cells. It is anticipated that this eco-friendly fluorescence probe may be widely used in monitoring food quality and as a fluorescent-imaging agent in pharmaceutical analysis.

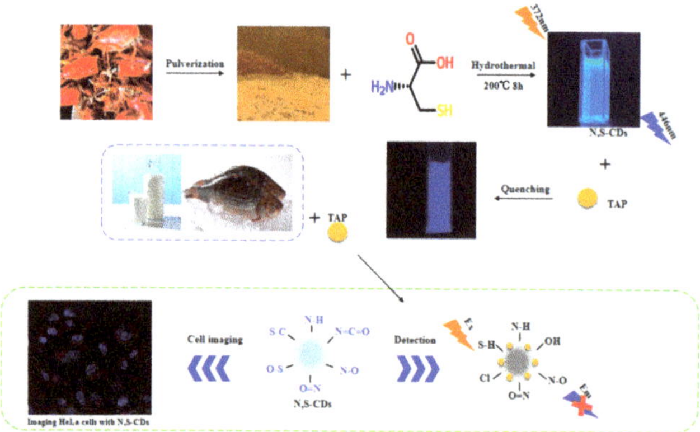

Figure 1. Schematic illustration of the synthesis of N, S-CDs and its application.

2. Materials and Methods

2.1. Materials

Crayfish shells were collected from restaurant waste in Changsha, China. The reference pharmaceuticals thiamphenicol (TAP), florfenicol (FF), chloramphenicol (CAP), amoxicillin (AMX), sulfonamide (SDI), penicillin (PG), doxycycline hydrochloride (DOX), melamine (MEL) ($\geq$98%), 2-(4-Amidinophenyl)-6-indolecarbamidine dihydrochloride (DAPI), and dimethyl sulfoxide (DMSO) were provided by Shanghai yuan ye Bio-Technology Co., Ltd. (Shanghai, China). L-cysteine was purchased from Chengdu Kelon Chemical Reagent Co., Ltd. (Chengdu, China). Britton–Robinson (BR) buffer solutions were prepared from H_3PO_4–HAc–H_3BO_3 mixed-acid solution. All the reagents were of analytical grade and used without further purification. Hela cell lines were obtained from Shanghai Binxin Bio-Technology Co., Ltd. (Shanghai, China). Ultrapure water (18.2 MΩ·cm) was produced by a Milli-Q ultrapure water system (Millipore, Burlington, MA, USA). Milk and fish samples were obtained from the Ya'an Jixuan supermarket (Ya'an, China).

2.2. Synthesis of N, S-CDs

N, S-doped CDs were prepared by hydrothermal treatment of crayfish shells. Crayfish shells (2 g) were pulverized and dispersed in 60 mL of ultrapure water, followed by the addition of L-cysteine (1 g). The mixture was transferred to a reaction kettle and then heated in an oven (200 °C, 8 h). After cooling to room temperature, the brownish dispersion was centrifuged at 12,000 rpm for 5 min to remove the precipitate. The supernatant was then collected by filtration (0.22 μm) and dialyzed (MW, 1000 Da) for 2 days to purify the N, S-CDs solution. Lastly, the resulting products were dried by lyophilization (−80 °C) to obtain N, S-CDs powder. Then, it was dispersed in water for further characterization and application.

2.3. Characterization of N, S-CDs

The morphologies and structures of N, S-CDs were explored by transmission electron microscopy (TEM) and high-resolution TEM (HRTEM) (JEOL-2100F electron microscope JEOL, Tokyo, Japan). The FTIR spectrum in KBr was acquired on a NICOLEF IS 10 FTIR apparatus (Thermo Fisher Scientific Co., Ltd., Waltham, MA, USA). The phase and elemental analyses were performed on a D/Max-2500 X-ray diffractometer (XRD, Shimadzu, Kyoto, Japan) and X-ray photoelectron spectroscope (XPS, ESCALAB250, Thermo, Waltham, MA, USA). The PL spectra and UV–vis absorption spectra were measured on a Varioskan Flash multimode reader (Thermo Fisher Scientific Co., Ltd., Waltham, MA, USA). Fluorescence lifetime measurements were performed with an Edinburgh FLSP920 time-corrected single photon-counting system. The cell imaging was taken by laser confocal fluorescence microscope (TCS SP5 II, Beijing Zhongyi Optical Technology Development Co., Ltd., Beijing, China). The Abs-QY of N, S-CDs were measured through a calibrated integrating sphere (IS) in the FLSP920 spectrometer (Edinburgh Instruments Ltd., Livingston, Scotland, UK). The Abs-QY value was calculated [25] as (Equation (1)):

$$\eta = \frac{Eb - Ea}{Sa - Sb} \quad (1)$$

where η is the absolute fluorescence quantum yield, the indices "*a*" and "*b*" referred to the reference sample (deionized water) and carbon dots aqueous solution separately, and *Ea*, *Eb*, *Sa*, and *Sb* referred to the integral of the scans.

2.4. Sensing Approaches for TAP Determination

The obtained N, S-CDs were mixed with BR buffer solutions (volume ratio: 2:1), and then, different concentrations of TAP (50–300 $\mu g \cdot L^{-1}$, 20–50 $\mu g \cdot L^{-1}$) were added into the mixture to form the N, S-CDs-TAP system (volume ratio of N, S-CDs, BR buffer, and TAP: 2:1:2), followed by vortexing the mixture for 30 s. The PL intensities were recorded at a $\lambda_{ex}/\lambda_{em}$ of 376/446 nm, and a linear standard curve for TAP sensing by N, S-CDs was established by the ratio of the fluorescence intensity of probes in the absence (F_0) and presence (F) of different concentrations of TAP. The selectivity of the detection system was investigated. Specifically, various representative antibiotics (FF, CAP, AMX, SDI, PG, DOX, and MEL) at the same concentration were added into the system instead of TAP, and the fluorescence intensity ratio were calculated under the same conditions ($\lambda_{ex}/\lambda_{em}$ of 376/446 nm). Interference experiments were performed by directly adding some biomolecules and metal ions (e.g., Ca^{2+}, Mg^{2+}, Na^{+}, Cl^{-}, K^{+}, sucrose, glucose, fructose, urea, and L-ascorbic acid) of high concentration (100–5000 $mg \cdot L^{-1}$), seen commonly in the food matrix, to the N, S-CDs-TAP system.

2.5. Analysis of Real Food Samples

Herein, the fish and milk real food samples were prepared and tested following a standard procedure to evaluate the eco-friendly probe's applicability in detecting TAP in foods [26,27]. Briefly, 2 g of spiked fish and milk samples (1 and 2 and 3 times that of the MRL value, respectively) [28] were placed in 50 mL centrifuge tubes, and 20 mL of ethyl acetate was added followed by thorough mixing. After centrifuging at 8000 rpm for 5 min to remove the protein, the extraction process was repeated twice. The supernatant of fish and milk samples were concentrated by nitrogen flow at 40 °C and re-dissolved in 5.0 mL ultrapure water. In the next step, this suspension was filtered through a 0.22 μm membrane filter. BR buffer and N, S-CDs solution were added into the solution and then vortexed for 3 min to obtain the sample ready for analysis. All the experiments were performed in triplicate.

2.6. Cellular Toxicity Test

The cytotoxicities of the N, S-CDs were evaluated on Hela cells by using colorimetric MTT assays [29]. Firstly, the Hela cells were cultured in the DMEM medium containing 10% fetal bovine serum to 7.5×10^4 $cells \cdot mL^{-1}$. Then, the as-obtained Hela cells were taken out as aliquot in 96-well plates containing N, S-CDs with different concentrations (0−550 $\mu g \cdot mL^{-1}$) and incubated for 24 h in an incubator. After 72 h of incubation, the cells were lysed with dimethyl sulfoxide (DMSO), and the absorbance of MTT at 570 nm optical density of 96-well plate was counted with microplate reader (Thermo Fisher Scientific

Co., Ltd., Waltham, MA, USA). The cell viabilities were calculated according to following formula (Equation (2)):

$$cell\ viability\% = \left(\frac{OD_{sample}}{OD_{blank}}\right) \times 100\% \quad (2)$$

where OD_{blank} is obtained in the absence of the sample, and OD_{sample} is obtained in the presence of N, S-CDs sample.

2.7. Cellular Imaging and Intracellular Detection of TAP in Hela Cells

The potential bioimaging application of the N, S-CDs was evaluated against the Hela cells through a procedure reported previously in the literature [14,30]. Firstly, the Hela cells were collected in the logarithmic growth phase and adjusted to 7.5×10^4 cells·mL^{-1} in DMEM medium with fetal bovine serum (10%). Then, they were seeded on 22 × 24 mm with sterile cell slides and incubated overnight at 37 °C under 5% CO_2. Afterward, discarding the supernatant, the culture medium was replaced with N, S-CDs containing serum-free medium for 24 h. The cells in coverslips were washed three times with PBS buffer, and 4% PFA was fixed for 15 min at 4 °C and washed three times with PBS. Lastly, the cells were stained for 15 min at room temperature using TAP or DAPI, followed by washing three times with PBS. Fluorescence imaging of the cells was observed using a laser confocal microscope. Each variable was assessed in triplicate.

3. Results and Discussion

3.1. Characterization of N, S-CDs

The synthesis mechanism diagram of N, S-CDs is shown in Figure 1. As shown in Figure 2a, the TEM results showed that N, S-CDs were homogeneous in size and morphology, without any obvious aggregation. A clear lattice fringe with a lattice spacing of 0.23 nm is displayed in the inset of Figure 2a, which can be interpreted as the (100) planes diffraction of the graphitic carbon framework [31]. It reveals the easy carbonization of the crayfish shell waste precursors to form a graphitic microstructure within the N, S-CDs. The average particle size diameter of carbon dots were generated by ImageJ software (Figure 2b). It can be observed that the average diameter of N, S-CD obtained in this study is 2.38 ± 0.14 nm.

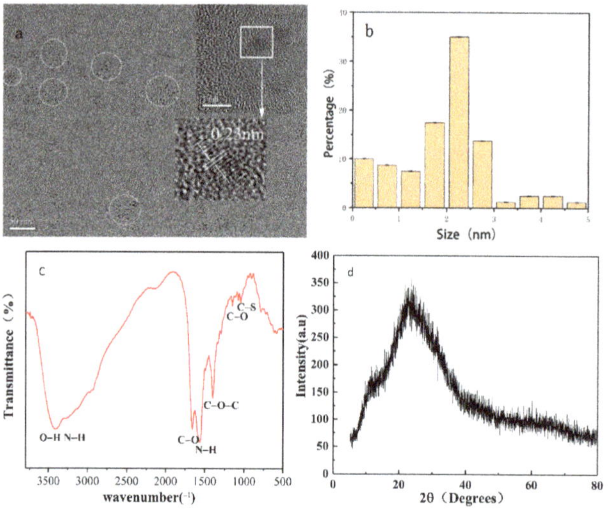

Figure 2. (**a**) The TEM imaging of the N, S-CDs; inset: the HRTEM imaging of the N, S-CDs; (**b**) the size distribution of the N, S-CDs; and (**c**) the FT-IR (**d**) and XRD of the N, S-CDs.

In the spectrum shown in Figure 2c, the broad absorption band in the region of 3000–3404 cm^{-1} can be assigned to O-H and N-H bending vibrations. The characteristic absorption peaks at 1656 and 1562 cm^{-1} could put down to the stretching vibration of C=O and N=O groups, respectively, and the 1397 cm^{-1} peak can be confirmed as C-N and -COO stretching bands [15,32]. Moreover, the C-O and C-S stretching bands [33] were situated at 1050 cm^{-1} and 1153 cm^{-1}. The peaks located at 1050 cm^{-1} and 787 cm^{-1} correspond to the C-O stretching vibrations and in-plane bending of the C-O bond [29], respectively. Lastly, a strong N-H stretching vibration [34] was observed at 669 cm^{-1}. Consequently, these FTIR spectral data demonstrated the existence of C=O, -NH_2, -OH, and C-S and C-N functional groups on the surface of N, S-CDs. It is apparent from Figure 2d that the XRD pattern of N, S-CDs displayed a typical wide peak at about 23.6°, which can be interpreted as the (002) planes of the graphitic carbon structure, confirming the amorphous carbon structure of N, S-CDs [21]. This observation may be attributed to the rich nitrogen, sulfur, oxygen, and other functional groups introduced [34] by the N, S-CDs sample. The surface functionalization process of N, S-CDs was further studied by XPS spectra. Four typical peaks at 170.1 eV, 289.9 eV, 400.8 eV, and 535.2 eV, as shown in Figure 3a, can be assigned as corresponding to S2p, C1s, N1s, and O1s, respectively [29]. Figure 3b demonstrated the four states of carbon atoms were C=C/C-C (284.4 eV), C-N/C-S (285.3 eV), C-O (286.6 eV), and C=N/N=O (288.2 eV), respectively [33]. The N1s XPS spectrum displayed three peaks at N-$(H)_3$ (399.5 eV), O=C-N (400.3 eV) [18], and N-H (401.2 eV) (Figure 3c). Two peaks located at 531.3 eV and 532.8 eV were observed in the O1s XPS spectrum (Figure 3d), which can be attributed to C=O and C-O-C/C-OH functional groups [32]. The high-resolution spectrum of S2p could be divided into three main peaks (Figure 3e), wherein the peaks at 163.5 eV and 164.5 eV were assigned to S2p3/2 and S2p1/2 of the C-S. The emergence of a peak at 168.1 eV could be attributed to the induction of sulfate or sulfonate (-C-SO_3^-) [15]. The above results indicated graphitic nature in carbon dots structure and verified the successful synthesis of N, S-CDs.

3.2. Optical Properties of N, S-CDs

According to the Figure 4a, a typical absorption peaks at 265 nm could be interpreted as the π-π* (C=C or C=O) [35] and n-π* (C=N or C-OH) electronic transitions. Studies have shown that doping properly could cause the change of local surface state and the recombination of excited-state energy, promoting the strong fluorescence emission of N, S-CDs in the form of main chromophore, which caused these characteristic absorptions to occur [31,36]. It can be found that the fluorescence intensity of N, S-CDs increased initially and then decreased, increasing the excitation wavelength (Figure 4b), and when excited at 372 nm, the N, S-CDs reflects the strongest blue fluorescence (inset of Figure 4a) near the emission peak of 446 nm. It is noteworthy that the positions of the emission peaks did not change significantly; this phenomenon exhibits that the PL emission of N, S-CDs does not depend on excitation, avoiding the interference of different excitations on the peak position of the sample emission [37], to achieve more accurate detection in complex food samples. Quantum yield is crucial for analyzing the properties of luminescent materials, and different from the relative quantum yield, the absolute quantum yield (Abs-QY) can remove the need for a known luminescence standard as a reference, thus avoiding obvious experimental system errors and accidental errors, and significantly simplifies the measurement process [38,39]. The Abs-QY of CDs and N, S-CDs were calculated to be 18.6% and 50.2%, respectively, which shows that nitrogen-sulfur doping significantly improves the absolute quantum yield of fluorescent probes. Compared with the CDs obtained by green synthesis in other reports, the QY of products in this research increased by two times [14,40]. Additionally, the stability of N, S-CDs in fluorescence and storage was also evaluated.

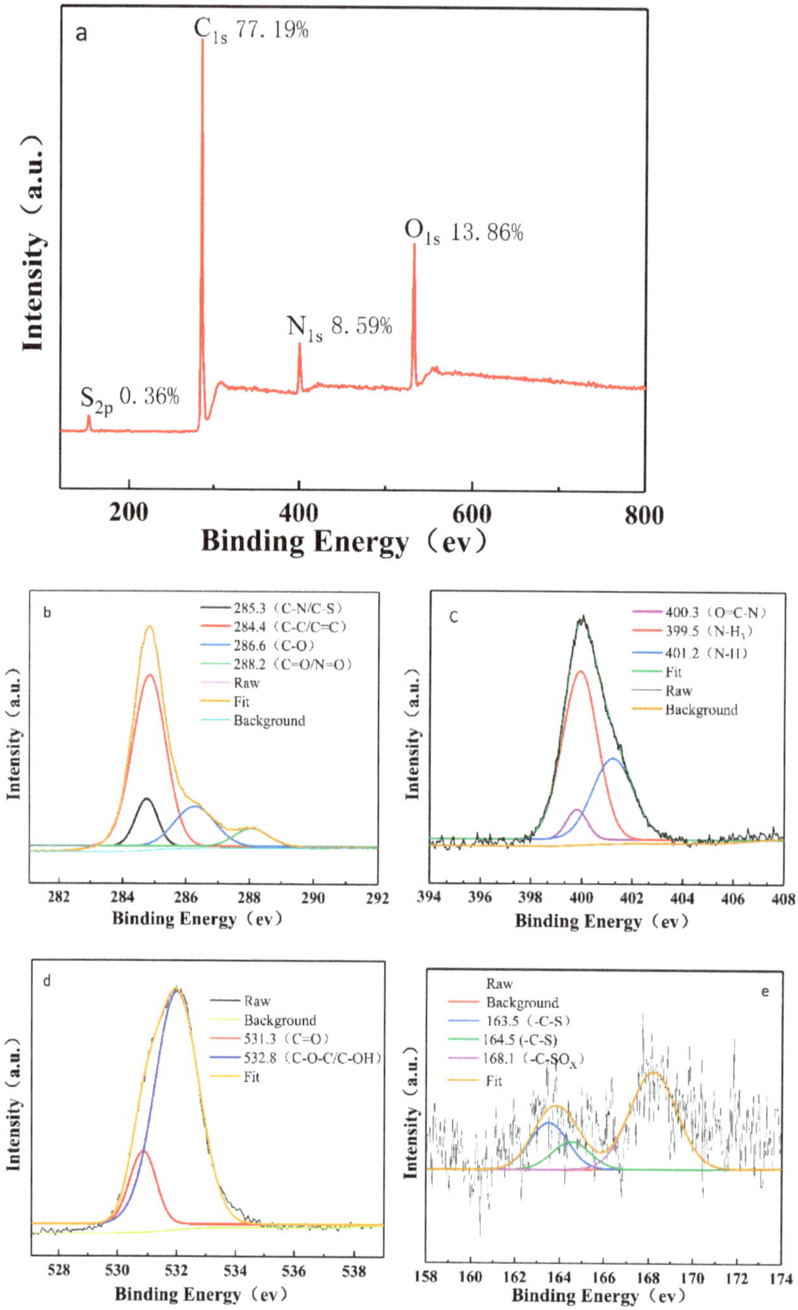

Figure 3. (**a**) The full-scan XPS spectrum of the N, S-CDs and (**b**) the high-resolution XPS of C1s, (**c**) N1s, (**d**) O1s, and (**e**) S2p.

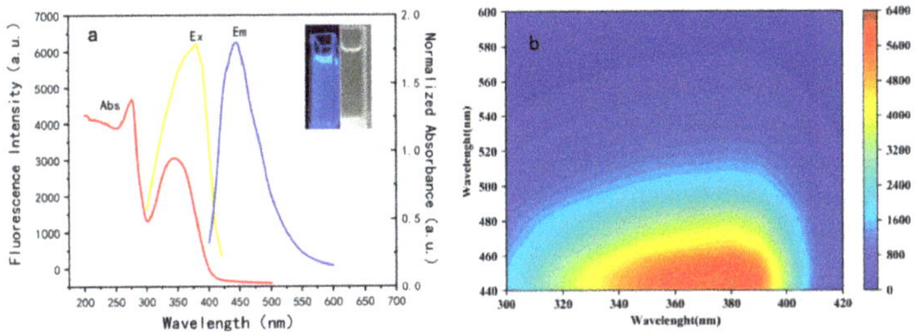

Figure 4. (**a**) Absorption and PL emission spectra (λ_{ex} = 372 nm) of the N, S-CDs; (**b**) PL emission spectra of the N, S-CDs at increasing excitation wavelengths from 300 nm to 420 nm.

3.3. Stability of the N, S-CDs

Figure 5a exhibits the emission behavior of N, S-CDs at different pH values (pH = 3–11). The PL intensity increased initially with an increase in the pH and then remained fairly constant in alkaline conditions. It shows a maximum value at pH = 8 (Ex = 372 nm, Em = 446 nm), which is believed to due to the presence of protonation and deprotonation. As shown in Figure 5b, the N, S-CDs possess excellent photoluminescence stability in the solutions at different ionic strengths (according to the concentration of NaCl), indicating that the N, S-CDs have high salt tolerance. At the same time, it could be observed that there is no substantial alteration in the fluorescence intensity with the storage time in 80 days (Figure 5c), which demonstrates that N, S-CDs have excellent stability for practical applications. Moreover, the proper concentration of N, S-CDs is crucial to the sensitivity of the fluorescent probe. According to reports, excessively high concentration of quantum dots does cause poor dispersion in aqueous solutions due to the generation of numerous nanoparticle clusters, and its non-radioactive energy transfer will also reduce the sensitivity of the method [41]. Therefore, the concentration of N, S-CDs in the range of 0.25–3.5 mg·mL^{-1} was also investigated (Figure 5d). With the increase in the concentration of N, S-CDs, the fluorescence signal showed an initial trend of rising and then falling and reached the maximum fluorescence intensity at 1.75 mg·mL^{-1}.

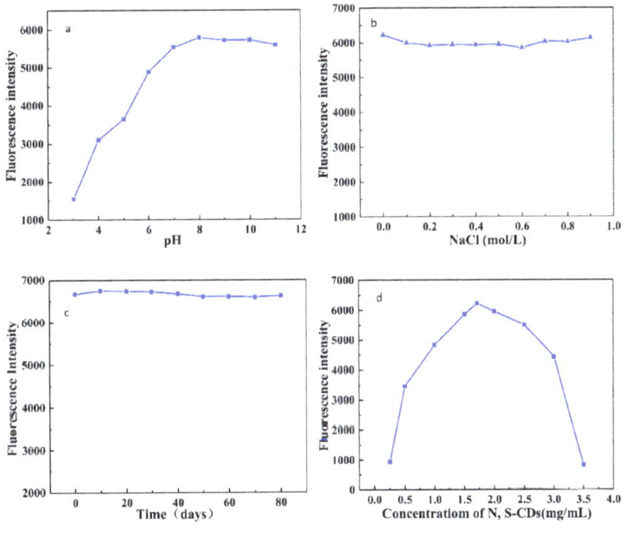

Figure 5. (**a**) Effect of pH on the PL intensities of the N, S-CDs; (**b**) effect of NaCl concentration on the PL intensities of the N, S-CDs; (**c**) effect of storage time on the PL intensities of the N, S-CDs; (**d**) effect of N, S-CDs concentration on the PL intensities of the N, S-CDs.

3.4. Optimization of the Sensing Conditions

pH has a significant influence on the fluorescence intensity of carbon dots and their interactions with target analytes. Figure 6a depicts the change in the quenching efficiency $(F_0 - F)/F_0$ of fluorescent probes on interaction with TAP in the pH range of 3–11. The PL intensities of N, S-CDs in the absence and presence of TAP are denoted by F_0 and F, respectively. The fluorescence intensity ratio $(F_0 - F)/F_0$ was the strongest at pH = 8, but at the same time, it remained nearly constant over a wide pH range. Therefore, the detection system is not restricted by different acid-base environments, which is highly advantageous for the application of probes to the food testing and biomedical industries. Since the probe shows a larger quenching efficiency at pH 8 than at other pH ranges, the BR buffer with a pH value of 8 was chosen as the solvent for the detection of TAP. As shown in Figure 6b, the effect of N, S-CDs concentration on quenching efficiency was also investigated. With the increase in the concentration of N, S-CDs, $(F_0 - F)/F_0$ shows an upward trend, reaching the maximum at a concentration of 0.5 $mg{\cdot}mL^{-1}$, and the continuous increase in the probe concentration leads to an unstable trend where the ratio of quenching repeatedly decreases and increases. As mentioned above, the excessive concentration of N, S-CDs produces too many nanoparticle clusters, which not only reduces the dispersibility of the CDs in the solution but also is not conducive to full interaction with the target. As such, 0.5 $mg{\cdot}mL^{-1}$ was chosen as the optimal concentration of N, S-CDs for the detection system. It is obvious from Figure 6c that after adding N, S-CDs to the TAP solution, the interaction of TAP with N, S-CDs occurred instantaneously. The fluorescence intensity ratio of the system reached an equilibrium within 5 min and was stable for at least 1.5 h, displaying the advantage of the probe for rapid detection. According to the analysis of optimization experiments, the reaction time, pH, and concentration of N, S-CDs, were determined to be 5 min, pH 8, and 0.5 $mg{\cdot}mL^{-1}$, respectively, as the conditions for subsequent experimental analysis.

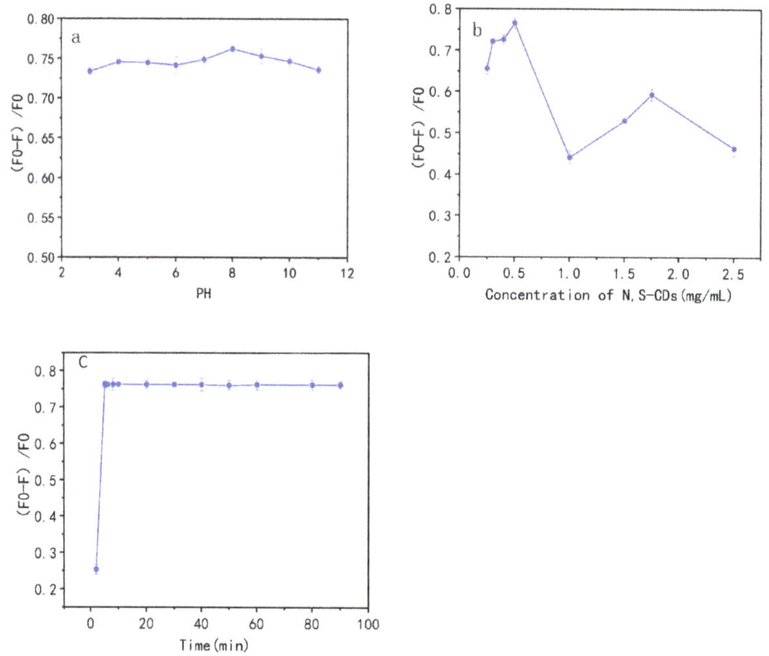

Figure 6. (**a**) The effect of pH on the fluorescent intensity of N, S-CDs-TAP system; (**b**) the effect of N, S-CDs concentration on the fluorescent intensity of N, S-CDs-TAP system; (**c**) the effect of incubation time on the fluorescence intensity of N, S-CDs.

3.5. Sensing of N, S-CDs to TAP

Under optimized conditions, this fluorescence probe was evaluated for TAP detection. As illustrated in Figure 7a,b, FL intensity of N, S-CDs significantly weakened as TAP concentration rose at the excitation wavelength of 372 nm, which showed good linearity at TAP concentrations from 50–300 μg·L^{-1}, with a linear equation of $(F_0 - F)/F_0 = 0.0012$CTAP + 0.03417 (R^2 = 0.9942), where F_0 and F are the fluorescence intensities of the detection system without and with TAP, respectively. Moreover, the limit of detection (LOD) was confirmed as 27.5 μg·L^{-1}, calculated according to the formula LOD = 3σ/S. As shown in Figure 7c,d, a linearity at TAP concentrations from 20–50 μg·L^{-1}, with a linear equation of $(F_0 - F)/F_0 = 0.0018$CTAP − 0.0137 (R^2 = 0.9846), the LOD was calculated as 11.12 μg·L^{-1}. All data indicate that the developed method can quantitatively detect the TAP and could achieve satisfactory results, such as detection limits below national standards.

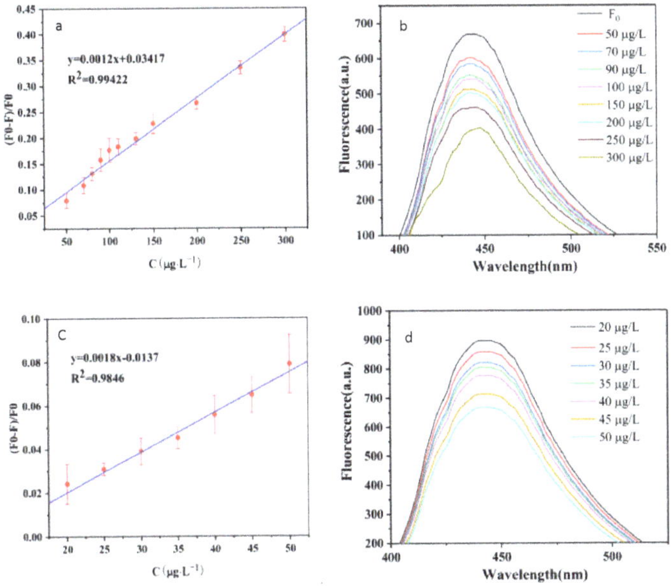

Figure 7. (**a**) The relationship between the fluorescence intensity and different concentrations of TAP (50–300 μg·L^{-1}); (**b**) fluorescence spectra of N, S-CDs after adding TAP (50–300 μg·L^{-1}); (**c**) the relationship between the fluorescence intensity and different concentrations of TAP (20–50 μg·L^{-1}); (**d**) fluorescence spectra of N, S-CDs after adding TAP (20–50 μg·L^{-1}).

3.6. Selectivity of N, S-CDs to TAP

The selectivity and interference of the N, S-CDs were evaluated to confirm the applicability of fluorescence sensors for food inspection. Several biological agents that are easily residual in food matrices were used to assess the selectivity of N, S-CDs. As shown in Figure S1, among these biological molecules, nonstructural analogues of thiamphenicol did not exhibit significant response effects (amoxicillin, sulfonamide, doxycycline, hydrochloride, melamine). However, chloramphenicol-based antibiotics, such as thiamphenicol, florfenicol, and chloramphenicol, exerted a significant response, and TAP had the highest fluorescence quenching efficiency. Afterward, we performed tests to verify the possibility of biological molecules that conventionally exist in milk samples interfering with the detection system by conducting interference test. Table S3 shows the investigated interference of the detection system. Little changes were observed in the FL intensity of N, S-CDs-TAP system under various interference factors such as high concentration 100–5000 mg·L^{-1}. The relative error was less than 2%.

3.7. Sensing Mechanism of N, S-CDs to TAP

To clearly understand the sensing mechanism of TAP detection, a series of researches were discussed, including zeta potential and variable temperature experimental (Ster-Volmer equation) analysis, UV–vis absorption spectrometry, and fluorescence lifetime analysis. The quenching process comprises energy transfer, ground-state complex formation, and collisions, etc. [34]. As shown by Figure S2A, compared to the absorption spectrum of N, S-CDs alone (curve b), the absorbance at 280 nm increased significantly after the addition of TAP quencher (curve a); moreover, a weak absorption peak appeared at 300 nm, which was attributed to the additional peak of the newly formed ground-state complex between the fluorophore and the quench agent through π–π^* stacking. Besides, the UV–vis spectrum of TAP exhibited a wide absorption peak (curve c) at around 220–300 nm. There was no overlap between N, S-CDs emission and TAP absorption; thus, inner filter effect (IFE) cannot be explained as the quenching mechanism of the detection process [29,42]. There should have been some interactions between N, S-CDs probes and TAP molecules. It is assumed that hydroxyl and halogen atoms of TAP form ground-state complexes such as amino and carboxyl groups or sulfhydryl groups with nitrogen-hydrogen bonds and carbonyl groups of N, S-CDs, which caused aggregation and hence reduced the π-π^* excitation in the N, S-CDs, finally resulting in the quenching of the fluorescence. Generally, the understanding of luminescence (fluorescence) decay dynamics of excited states of carbon dots is significant for understanding its optical mechanism and catalytic properties [43]. The fluorescent lifetime can represent the information about the excited state of matter and the presence of a dynamic quencher to shorten the fluorescence lifetime of an excited molecule fluorophore. In contrast, the static quencher does not apparently change the fluorescence lifetime of a fluorophore [42]. The fluorescence decay curves of N, S-CDs with and without TAP were measured, as depicted in Figure S2B. The fluorescent lifetime of N, S-CDs was determined to be 1.72 ns, which turned out to be 1.58 ns after the addition of TAP. The almost constant values demonstrate that the interaction between TAP and N, S-CDs belongs to the category of a static quenching process. Dynamic quenching depends upon diffusion. When the temperature rises, the viscosity of the solution decreases and the movement of molecules accelerates the results in larger diffusion coefficients, and then, the bimolecular quenching constants are supposed to increase. On the contrary, the increase in temperature may cause the stability of the complex to decrease, thereby reducing the degree of static quenching. Comparing the Stern–Volmer quenching constant (KSV) according to the Stern–Volmer equation, the slope obtained for the data recorded at high temperature is lower than that at low temperature (Table S4), which suggests that the mechanism of TAP-induced quenching of N, S-CDs excludes the possibility of dynamic collisions but was attributed to the formation of complexes [11]. The zeta potential of N, S-CDs was measured to be nearly −17.4 mV, which was attributed to the large number of nitrogen hydrogen bond and carbonyl groups present on the surface of N, S-CDs. However, the zeta-potential dropped to −26.70 mV after adding TAP to the N, S-CDs solutions, showing that the hydroxyl and hydrogen bonds in TAP interact with N, S-CDs to generate complexes such as carboxyl and amine groups, which is further validation of the static quenching results [36].

3.8. Detection of Actual Food Samples

To further verify this probe's feasibility and applicability, the proposed method was applied to test TAP concentration in some real food samples such as milk and fish. The results are presented in Table 1.

Table 1. Determination of TAP in spiked real samples by N, S-CDs (n = 3).

Sample	Added (μg·L^{-1})	Found (μg·L^{-1})	Recovery	RSD (%)
Milk-1	0	nd [a]	-	-
Milk-2	50	49.20 ± 0.77	98.39%	1.54%
Milk-3	100	98.20 ± 1.04	98.20%	1.04%
Milk-4	150	149.00 ± 1.33	99.34%	0.89%
Fish-1	0	nd [a]	-	-
Fish-2	50	48.91 ± 0.20	97.30%	0.41%
Fish-3	100	96.16 ± 0.67	97.86%	0.69%
Fish-4	150	148.64 ± 1.99	97.92%	1.34%

nd [a] defined as "not detected".

As Table 1 shows, the obtained recoveries of TAP from spiked milk and fish samples ranged from 97.30 to 99.34%, with the relative standard deviation (RSD) values all being below 2% (n = 3). The data reveal that the environment-friendly fluorescent nanoprobe based on N, S-CDs proposed in this research could be a practical approach for accurate TAP detection in complex food. The analytical characteristics of the other reported methods are listed in Table 2. Compared with ELISA techniques, the probes developed in this work do not require using susceptible and highly expensive bio-recognition molecules (such as antibodies, aptamers, or enzymes) and has a comparable linear range (50–300/20–50 μg·L^{-1}). Furthermore, the LOD (27.5/11.12 μg·L^{-1}) values obtained by the present method are better than UV–vis spectrophotometric methods. Although better detection limits were achieved with chromatographic and electrochemical methods, these require elaborate and sophisticated instrumentation and labor-intensive operation. Importantly, as for the method our developed, the fast and effective pretreatment process can effectively shorten the time to detect TAP in real foods (45 min). Meanwhile, compared with other CDs fluorescent probes [31,35], this probe has faster fluorescence response time (5 min) and benefits from its excellent fluorescence properties such as high quantum yield and stable fluorescence lifetime.

Table 2. Comparison of different methods for the detection of TAP.

Technique	Pretreatments	Detection Time	Linear Range (μg·L^{-1})	LOD (μg·L^{-1})	Recovery (%)	Ref.
ELISA	Prepare and screen the polyclonal antibodies	>70 min	0.41–11.2	0.15	77.2–116.0	[7]
UV spectrophotometric	Purification	55 min	5000–25,000	590	99.26–100.56	[8]
HPLC	Thin-layer chromatography (TLC) purification	>1 h	50–10,000	20	82.0–114.9	[9]
HPLC-DAD	dispersive solid phase extraction (SPE)	2 h	5–150	0.48	81.7–97.5	[44]
UHPLC-MS	SPE	>1 h	1–200	0.5	87.6–102.7	[1]
HPLC-ESI/MS	Accelerated solvent extraction (ASE)	>1 h	1–150	0.4	88.3–107	[10]
Fe3O4 @ SiO2-MIPS	Purification and separation by organic solvents	90 min	100–300	10	92.6–95.6	[5]
CNTs-AuNPs electrochemical sensor	Modification of screen printed electrode	>30 min	35.5–10,650	1.07	94.0–97.0	[45]
N, S-CDs	Purification	45 min	50–300	27.5	97.30–99.34	This work

3.9. Cytotoxicity and Cellular Imaging

Figure S3 exhibits the extremely low cytotoxicity of N, S-CDs, and the cell viability of Hela cells can still reach 91.76% even at the incubation of 550 μg·mL^{-1} of N, S-CDs. Because of the low toxicity assay and outstanding optical merits of N, S-CDs mentioned above, this probe was used as a means for the fluorometric detection of TAP in living Hela cells. The cells were treated previously with N, S-CDs. It is clear from Figure 6A–D that the non-treated cells showed almost no background fluorescence, while the cells with N, S-CDs showed a light fluorescence; the results reveal that N, S-CDs can penetrate the cell membrane without restriction and be positioned in the cytoplasm. At the same time, it makes the intracellular biosensing feasible. From Figure 8E–H, it is noticeable that compared to the control cells, those supplemented with TAP (50, 100, 250 $\mu g \cdot mL^{-1}$) have a significant decrease in fluorescence intensity. This highlights the advantages and great potential of N, S-CDs as a novel bioluminescent probe for monitoring TAP in living cells.

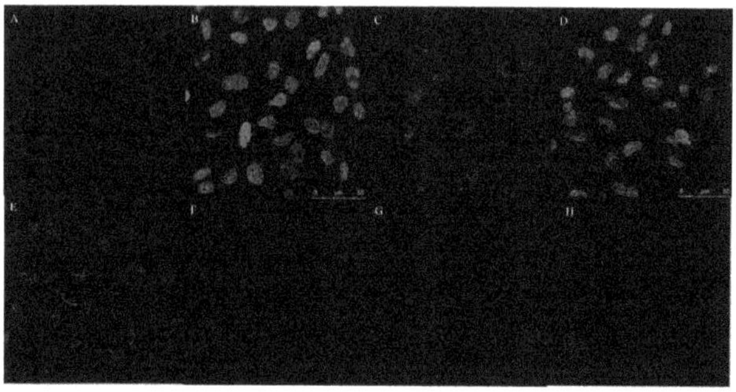

Figure 8. The confocal fluorescence images of Hela cells incubated with N, S-CDs or DAPI; (**A**) control group; (**B**) DAPI; (**C**) N, S-CDs; (**D**) overlap of N, S-CDs, and DAPI; (**E**) the confocal fluorescence images of Hela cells labeled with N, S-CDs (control group); (**F**) confocal fluorescence images of Hela cells labeled with N, S-CDs and TAP (50 $\mu g \cdot mL^{-1}$); (**G**) confocal fluorescence images of Hela cells labeled with N, S-CDs, and TAP (100 $\mu g \cdot mL^{-1}$); and (**H**) confocal fluorescence images of Hela cells labeled with N, S-CDs and TAP (250 $\mu g \cdot mL^{-1}$).

4. Conclusions

In summary, a novel, facile, and low-cost fluorescent probe N, S-CDs prepared from crayfish shell waste precursors was utilized for the selective detection of TAP and its analogues. The probe exhibited high fluorescence activity with Abs-QY of 50.2% and stable fluorescence lifetime and displayed outstanding properties in terms of detecting the TAP in food samples, for example, wide linear range (20–300 $\mu g \cdot L^{-1}$), high selectivity, and good limits of detection (LOD = 11.12 $\mu g \cdot L^{-1}$). Specifically, the rapid detection time was within 45 min compared with other methods and the recoveries varied from 97.30% to 99.34% with the RSDs < 2%. We concluded that the formation of a non-fluorescent complex from N, S-CDs-TAP system is most likely accountable for the quenching phenomenon. Furthermore, the results proved that the probes made of environmentally friendly materials exhibited significantly low toxicity and FL intensity, which further demonstrated the potential of intracellular detection of TAP and cell imaging. More effort should be devoted to the commercialization of the N, S-CDs due to their low cost, environmentally friendly nature, and simple and rapid detection method, which provides new insights into food quality monitoring and biomedical control fields.

5. Patents

A patent for invention resulted from the work reported in this manuscript: A fluorescent probe of carbon quantum dots and a method for detecting the content of thiamphenicol (Patent number: ZL 2021 1 0156639.9).

Supplementary Materials: The following supporting information can be downloaded at: https://www.mdpi.com/article/10.3390/foods11162414/s1, Figure S1: selectivity of the N, S-CDs to different kinds of antibiotics; Figure S2: Supplementary experimental content: Sensing mechanism of N, S-CDs to TAP; Figure S3: Cytotoxicity of the N, S-CDs toward Hela cells; Table S1: Comparison of N, S-CDs prepared from different precursor; Table S2: Comparison of fluorescence response of N, S-CDs prepared from different food wastes and dopant to TAP; Table S3: Effect of interfering substances of C-dots-TAP solution system; Table S4: Comparison temperatures of different research for the calculated KSV values; Table S5. Specification information for N, S-CDs. References [46–48] are cited in the supplementary materials.

Author Contributions: Conceptualization, S.C. and W.O.; methodology, W.O.; validation, S.C., W.O. and Y.Z.; formal analysis, W.O.; investigation, W.O. and Y.Z.; resources, S.C.; data curation, Y.Z.; writing—original draft preparation, W.O.; writing—review and editing, L.H., L.Z. and X.A.; supervision, S.L. and Y.Y.; visualization, W.O. and J.L.; project administration, W.O. and S.C.; funding acquisition, S.C. All authors have read and agreed to the published version of the manuscript.

Funding: This work was funded by Sichuan Science and Technology Program (No. 2020ZHCG0044, No. 2022ZDZX0017 and No. 2021YFH0192), Project Technology Innovation Research and Development of Chengdu Science and Technology Bureau (2022-YF05-00487-SN).

Institutional Review Board Statement: Not applicable.

Informed Consent Statement: Not applicable.

Data Availability Statement: Data are contained within the article.

Acknowledgments: The authors thank Tianyi Zhang for helpful discussion and help and thank Tianyi Zhang for providing the excellent technical assistance, and some of the experiments were performed in their laboratory.

Conflicts of Interest: The authors declare no conflict of interest.

References

1. Li, J.; Gong, J.; Yuan, H.; Xiao, G.; Wang, H.; Sun, L.; Qiu, H.; Chu, B.; Fang, R. Determination of Chloramphenicol, Thiamphenicol and Florfenicol in Chinese Gelatin Medicines using Dispersive Solid-Phase Extraction Coupled with Ultra High-Performance Liquid Chromatography-Mass Spectrometry. *J. Chromatogr. Sci.* **2020**, *58*, 471–476. [CrossRef] [PubMed]
2. Impens, S.; Reybroeck, W.; Vercammen, J.; Courtheyn, D.; Ooghe, S.; De Wasch, K.; Smedts, W.; De Brabander, H. Screening and confirmation of chloramphenicol in shrimp tissue using ELISA in combination with GC–MS2 and LC–MS2. *Anal. Chim. Acta* **2003**, *483*, 153–163. [CrossRef]
3. Thiam, A.; Salazar, R.; Brillas, E.; Sires, I. In-situ dosage of Fe^{2+} catalyst using natural pyrite for thiamphenicol mineralization by photoelectro-Fenton process. *J. Environ. Manag.* **2020**, *270*, 110835. [CrossRef] [PubMed]
4. Samsonova, J.V.; Cannavan, A.; Elliott, C.T. A Critical Review of Screening Methods for the Detection of Chloramphenicol, Thiamphenicol, and Florfenicol Residues in Foodstuffs. *Crit. Rev. Anal. Chem.* **2012**, *42*, 50–78. [CrossRef]
5. Liu, H.; Zhou, Y.; Qi, Y.; Sun, Z.; Gong, B. Preparation of thiamphenicol magnetic surface molecularly imprinted polymers for its selective recognition of thiamphenicol in milk samples. *J. Liq. Chromatogr. Relat. Technol.* **2018**, *41*, 868–879. [CrossRef]
6. Commission Regulation (EU) No 37/2010 of 22 December 2009 on Pharmacologically Active Substances and Their Classification Regarding Maximum Residue Limits in Foodstuffs of Animal Origin. *Off. J. Eur. Communities* **2010**, *15*. Available online: http://data.europa.eu/eli/reg/2010/37(1)/oj (accessed on 26 May 2021).
7. Li, R.; Lin, Z.-J.; Yang, J.-Y.; Xu, Z.-L.; Wang, H.; Lei, H.-T.; Sun, Y.-M.; Shen, Y.-D. An Indirect Competitive Enzyme-linked Immunosorbent Assay for Simultaneous Determination of Florfenicol and Thiamphenicol in Animal Meat and Urine. *Chin. J. Anal. Chem.* **2018**, *46*, 1321–1328. [CrossRef]
8. Martins, Y.; Oliveira, D. Development and Validation of an UV Spectrophotometric Method for Determination of Thiamphenicol in Dosage Form. *J. Appl. Spectrosc.* **2019**, *86*, 629–635. [CrossRef]
9. Wang, J.; Yang, J.; Qian, M.; Zhou, D.; Ke, X.; Sun, G.; Yang, B. A TLC-HPLC Method for Determination of Thiamphenicol in Pig, Chicken, and Fish Feedstuffs. *Food Anal. Methods* **2018**, *11*, 3103–3112. [CrossRef]
10. Wang, B.; Zhao, X.; Xie, X.; Xie, K.; Zhang, G.; Zhang, T. Development of an Accelerated Solvent Extraction Approach for Quantitative Analysis of Chloramphenicol, Thiamphenicol, Florfenicol, and Florfenicol Amine in Poultry Eggs. *Food Anal. Methods* **2019**, *12*, 1705–1714. [CrossRef]
11. Tsai, H.; Hu, H.C.; Hsieh, C.C.; Lu, Y.H.; Chen, C.H.; Fuh, C.B. Fluorescence studies of the interaction between chloramphenicol and nitrogen-doped graphene quantum dots and determination of chloramphenicol in chicken feed. *J. Chin. Chem. Soc.* **2019**, *67*, 152–159. [CrossRef]
12. Liu, Z.; Wang, X.; Hou, J.; Huo, D.; Hou, C. A Simple and Rapid Phosphorescence Probe Based on Mn-Doped ZnS Quantum Dots for Chloramphenicol Detection. *Nano* **2019**, *14*, 1950144. [CrossRef]
13. Fan, H.; Zhang, M.; Bhandari, B.; Yang, C.-H. Food waste as a carbon source in carbon quantum dots technology and their applications in food safety detection. *Trends Food Sci. Technol.* **2020**, *95*, 86–96. [CrossRef]
14. Chatzimitakos, T.; Kasouni, A.; Sygellou, L.; Avgeropoulos, A.; Troganis, A.; Stalikas, C. Two of a kind but different: Luminescent carbon quantum dots from Citrus peels for iron and tartrazine sensing and cell imaging. *Talanta* **2017**, *175*, 305–312. [CrossRef] [PubMed]
15. Chen, Y.; Wu, Y.; Weng, B.; Wang, B.; Li, C. Facile synthesis of nitrogen and sulfur co-doped carbon dots and application for Fe(III) ions detection and cell imaging. *Sens. Actuators B Chem.* **2016**, *223*, 689–696. [CrossRef]

16. Wei, J.-M.; Liu, B.-T.; Zhang, X.; Song, C.-C. One-pot synthesis of N, S co-doped photoluminescent carbon quantum dots for Hg^{2+} ion detection. *New Carbon Mater.* **2018**, *33*, 333–340. [CrossRef]
17. Wu, H.; Tong, C. Nitrogen- and Sulfur-Codoped Carbon Dots for Highly Selective and Sensitive Fluorescent Detection of Hg^{2+} Ions and Sulfide in Environmental Water Samples. *J. Agric. Food Chem.* **2019**, *67*, 2794–2800. [CrossRef]
18. Shi, W.; Guo, F.; Han, M.; Yuan, S.; Guan, W.; Li, H.; Huang, H.; Liu, Y.; Kang, Z. N,S co-doped carbon dots as a stable bio-imaging probe for detection of intracellular temperature and tetracycline. *J. Mater. Chem. B* **2017**, *5*, 3293–3299. [CrossRef]
19. Wang, J.; Wu, Z.; Chen, S.; Yuan, R.; Dong, L. A novel multifunctional fluorescent sensor based on N/S co-doped carbon dots for detecting Cr (VI) and toluene. *Microchem. J.* **2019**, *151*, 104246. [CrossRef]
20. Chen, J.; Liu, J.; Li, J.; Xu, L.; Qiao, Y. One-pot synthesis of nitrogen and sulfur co-doped carbon dots and its application for sensor and multicolor cellular imaging. *J. Colloid Interface Sci.* **2017**, *485*, 167–174. [CrossRef]
21. Ding, C.; Deng, Z.; Chen, J.; Jin, Y. One-step microwave synthesis of N, S co-doped carbon dots from 1,6-hexanediamine dihydrochloride for cell imaging and ion detection. *Colloids Surf. B Biointerfaces* **2020**, *189*, 110838. [CrossRef] [PubMed]
22. Wang, Y.; Kim, S.H.; Feng, L. Highly luminescent N, S- Co-doped carbon dots and their direct use as mercury(II) sensor. *Anal. Chim. Acta* **2015**, *890*, 134–142. [CrossRef] [PubMed]
23. Dong, Y.; Pang, H.; Yang, H.B.; Guo, C.; Shao, J.; Chi, Y.; Li, C.M.; Yu, T. Carbon-based dots co-doped with nitrogen and sulfur for high quantum yield and excitation-independent emission. *Angew. Chem. Int. Ed.* **2013**, *52*, 7800–7804. [CrossRef]
24. Zhao, X.R. Research on Proteinous Adhesive Made of Protein Isolated from Crayfish Shell Waste. *Adv. Mater. Res.* **2013**, *815*, 299–304. [CrossRef]
25. Liu, X.; Pang, J.; Xu, F.; Zhang, X. Simple Approach to Synthesize Amino-Functionalized Carbon Dots by Carbonization of Chitosan. *Sci. Rep.* **2016**, *6*, 31100. [CrossRef] [PubMed]
26. *GB 29689-2013*; National Food Safety Standard-Determination of Thiamphenicol Residues in Milk-HPLC Method. Ministry of Agriculture of the People's Republic of China: Beijing, China, 2013.
27. The Ministry of Agriculture of the People's Republic of China. Determination of Chloramphenicol, Thiamphenicol and Florfenicol Residues in Fishery Products Gas Chromatography Method (Announcement of the Ministry of Agriculture [13-2007] No. 958). Ministry of Agriculture: Beijing, China. 18 December 2006. Available online: http://jiuban.moa.gov.cn/nybgb/2008/deq/201806/t20180609_6151543.htm (accessed on 26 May 2021).
28. Jalili, R.; Khataee, A.; Rashidi, M.R.; Razmjou, A. Detection of penicillin G residues in milk based on dual-emission carbon dots and molecularly imprinted polymers. *Food Chem.* **2020**, *314*, 126172. [CrossRef] [PubMed]
29. Zhao, N.; Wang, Y.; Hou, S.; Zhao, L. Functionalized carbon quantum dots as fluorescent nanoprobe for determination of tetracyclines and cell imaging. *Mikrochim. Acta* **2020**, *187*, 351. [CrossRef]
30. Kascakova, S.; Maigre, L.; Chevalier, J.; Refregiers, M.; Pages, J.M. Antibiotic transport in resistant bacteria: Synchrotron UV fluorescence microscopy to determine antibiotic accumulation with single cell resolution. *PLoS ONE* **2012**, *7*, e38624. [CrossRef]
31. Zhang, Y.; Gao, Z.; Yang, X.; Chang, J.; Liu, Z.; Jiang, K. Fish-scale-derived carbon dots as efficient fluorescent nanoprobes for detection of ferric ions. *RSC Adv.* **2019**, *9*, 940–949. [CrossRef]
32. Liao, S.; Zhao, X.; Zhu, F.; Chen, M.; Wu, Z.; Song, X.; Yang, H.; Chen, X. Novel S, N-doped carbon quantum dot-based "off-on" fluorescent sensor for silver ion and cysteine. *Talanta* **2018**, *180*, 300–308. [CrossRef]
33. Gu, D.; Hong, L.; Zhang, L.; Liu, H.; Shang, S. Nitrogen and sulfur co-doped highly luminescent carbon dots for sensitive detection of Cd (II) ions and living cell imaging applications. *J. Photochem. Photobiol. B* **2018**, *186*, 144–151. [CrossRef] [PubMed]
34. Konar, S.; Samanta, D.; Mandal, S.; Das, S.; Mahto, M.K.; Shaw, M.; Mandal, M.; Pathak, A. Selective and sensitive detection of cinnamaldehyde by nitrogen and sulphur co-doped carbon dots: A detailed systematic study. *RSC Adv.* **2018**, *8*, 42361–42373. [CrossRef] [PubMed]
35. Gao, R.; Wu, Z.; Wang, L.; Liu, J.; Deng, Y.; Xiao, Z.; Fang, J.; Liang, Y. Green Preparation of Fluorescent Nitrogen-Doped Carbon Quantum Dots for Sensitive Detection of Oxytetracycline in Environmental Samples. *Nanomaterials* **2020**, *10*, 1561. [CrossRef] [PubMed]
36. Xu, S.; Ye, S.; Xu, Y.; Liu, F.; Zhou, Y.; Yang, Q.; Peng, H.; Xiong, H.; Zhang, Z. Microwave-assisted Synthesis of N,S-co-carbon Dots as Switch-on Fluorescent Sensor for Rapid and Sensitive Detection of Ascorbic Acid in Processed Fruit Juice. *Anal. Sci.* **2020**, *36*, 353–360. [CrossRef] [PubMed]
37. Zhang, T.; Dong, S.; Zhao, F.; Deng, M.; Fu, Y.; Lü, C. Tricolor emissive carbon dots for ultra-wide range pH test papers and bioimaging. *Sens. Actuators B Chem.* **2019**, *298*, 126869. [CrossRef]
38. Porres, L.; Holland, A.; Palsson, L.O.; Monkman, A.P.; Kemp, C.; Beeby, A. Absolute measurements of photoluminescence quantum yields of solutions using an integrating sphere. *J. Fluoresc.* **2006**, *16*, 267–272. [CrossRef]
39. Valenta, J. Photoluminescence of the integrating sphere walls, its influence on the absolute quantum yield measurements and correction methods. *AIP Adv.* **2018**, *8*, 105123. [CrossRef]
40. Shi, L.; Zhao, B.; Li, X.; Zhang, G.; Zhang, Y.; Dong, C.; Shuang, S. Green-fluorescent nitrogen-doped carbon nanodots for biological imaging and paper-based sensing. *Anal. Methods* **2017**, *9*, 2197–2204. [CrossRef]
41. Sadeghi, S.; Olieaei, S. Capped cadmium sulfide quantum dots with a new ionic liquid as a fluorescent probe for sensitive detection of florfenicol in meat samples. *Spectrochim. Acta A Mol. Biomol. Spectrosc.* **2019**, *223*, 117349. [CrossRef]
42. Hu, Q.; Gao, L.; Rao, S.Q.; Yang, Z.Q.; Li, T.; Gong, X. Nitrogen and chlorine dual-doped carbon nanodots for determination of curcumin in food matrix via inner filter effect. *Food Chem.* **2019**, *280*, 195–202. [CrossRef]

43. Röding, M.; Bradley, S.J.; Nydén, M.; Nann, T. Fluorescence Lifetime Analysis of Graphene Quantum Dots. *J. Phys. Chem. C* **2014**, *118*, 30282–30290. [CrossRef]
44. Patyra, E.; Kwiatek, K. HPLC-DAD analysis of florfenicol and thiamphenicol in medicated feedingstuffs. *Food Addit. Contam. Part A Chem. Anal. Control Expo. Risk Assess.* **2019**, *36*, 1184–1190. [CrossRef] [PubMed]
45. Muhammad, A.; Hajian, R.; Yusof, N.A.; Shams, N.; Abdullah, J.; Woi, P.M.; Garmestani, H. A screen printed carbon electrode modified with carbon nanotubes and gold nanoparticles as a sensitive electrochemical sensor for determination of thiamphenicol residue in milk. *RSC Adv.* **2018**, *8*, 2714–2722. [CrossRef] [PubMed]
46. Christopoulou, N.-M.; Kalogianni, D.P.; Christopoulos, T.K. Posidonia oceanica (Mediterranean tapeweed) leaf litter as a source of fluorescent carbon dot preparations. *Microchem. J.* **2021**, *161*, 105787. [CrossRef]
47. Hu, Q.; Paau, M.C.; Zhang, Y.; Gong, X.; Zhang, L.; Lu, D.; Liu, Y.; Liu, Q.; Yao, J.; Choi, M.M.F. Green synthesis of fluorescent nitrogen/sulfur-doped carbon dots and investigation of their properties by HPLC coupled with mass spectrometry. *RSC Adv.* **2014**, *4*, 18065–18073. [CrossRef]
48. Hu, Y.; Yang, J.; Tian, J.; Jia, L.; Yu, J.-S. Waste frying oil as a precursor for one-step synthesis of sulfur-doped carbon dots with pH-sensitive photoluminescence. *Carbon* **2014**, *77*, 775–782. [CrossRef]

foods

MDPI

Article

Squalene-Rich Amaranth Oil Pickering Emulsions Stabilized by Native α-Lactalbumin Nanoparticles

Andrea P. Cuevas-Gómez [1], Berenice González-Magallanes [1], Izlia J. Arroyo-Maya [2,*], Gustavo F. Gutiérrez-López [1], Maribel Cornejo-Mazón [3] and Humberto Hernández-Sánchez [1,*]

1 Departamento de Ingeniería Bioquímica, Escuela Nacional de Ciencias Biológicas, Instituto Politécnico Nacional, Mexico City C.P. 07738, Mexico; andreac327@gmail.com (A.P.C.-G.); bgm-1621@hotmail.com (B.G.-M.); gusfgl@gmail.com (G.F.G.-L.)

2 División de Ciencias Naturales e Ingeniería, Universidad Autónoma Metropolitana Cuajimalpa, Mexico City C.P. 05348, Mexico

3 Departamento de Biofísica, Escuela Nacional de Ciencias Biológicas, Instituto Politécnico Nacional, Mexico City C.P. 11340, Mexico; maribelpabe2@hotmail.com

* Correspondence: iarroyo@cua.uam.mx (I.J.A.-M.); hhernandezs@ipn.mx (H.H.-S.)

Abstract: The stabilization of Pickering emulsions by nanoparticles has drawn great interest in the field of food science and technology. In this study, α-Lactalbumin nanoparticles prepared by the desolvation and cross-linking method from protein solutions with initial pH values of 9 and 11 were used to stabilize squalene-rich amaranth oil Pickering o/w emulsions. The effect of different concentrations of nanoparticles on the size, size distribution, ζ potential, and emulsion stability was evaluated using dynamic light scattering, electron microscopy, and light backscattering. Dependence of the emulsions' droplet size on the nanoparticle concentration was observed, and the critical coverage ratio was reached when 5–10% nanoparticles concentration was used. Our findings suggest that α-LA nanoparticles at a 10% concentration can be used as novel stabilizers for Pickering emulsions to provide protection for beneficial lipophilic bioactive compounds. This is the first time that native α-LA nanoparticles have been used as stabilizers of Pickering emulsions.

Keywords: amaranth oil; Pickering emulsion; α-lactalbumin; nanoparticles

1. Introduction

Pickering emulsions are emulsions in which solid colloidal particles are used instead of molecular surfactants to convey kinetic stability [1]. Pickering emulsions are more stable than the "traditional" emulsions since the solid particles bind firmly to the oil-water interface, creating a stiff barrier against coalescence [2]. Natural macromolecules such as zein [3], soy [4] or whey proteins [5], and cellulose or starch particles [6] have been used as Pickering stabilizers for food applications due to their biocompatibility, biodegradability, and low cost [7,8]. Emulsions are adequate options for delivering different kinds of lipophilic bioactive compounds which are poorly soluble in water, chemically unstable, and sensitive to oxidation such as α-tocopherol [6], conjugated linoleic acid [7], polyphenols, and flavonoids [9]. Squalene was selected to be encapsulated in a Pickering emulsion since this compound has all the disadvantages mentioned above for hydrophobic compounds. The preparation of food-grade squalene-rich amaranth oil Pickering emulsions would provide a system containing a very valuable oil stabilized by edible solid protein particles which can be used for the formulation of functional foods (usually spread-like products) or dietary supplements.

Squalene is a natural linear triterpene, structurally similar to β-carotene, that can be found abundantly in shark liver oil, and to a lesser extent in a broad variety of vegetable oils such as amaranth, olive, and rice bran oils [10,11]. Squalene is well known for its key role as an intermediate in cholesterol synthesis [12]. Its bioactive properties have been widely studied. In vitro and in vivo studies revealed strong antioxidant properties which were able to reduce stress-induced intracellular reactive oxygen species (ROS). In the skin, squalene can act as an emollient and UV-protective compound. Additionally, squalene reduces pro-inflammatory cytokine secretion resulting in an overall protective role against uncontrolled inflammatory response [11]. There are also several studies indicating favorable effects of squalene-enriched diets on some cardiovascular pathologies. These effects included a decrease in plasma cholesterol and triglyceride levels, liver fat levels, and a reduction in systolic blood pressure [13].

Shark-derived squalene has been providing the global market needs since its initial discovery. Nevertheless, intensive fishing has had a damaging effect on oceanic ecosystems and endangered the populations of squalene-producing sharks [11]. The use of vegetable oils to obtain squalene has been considered then, as a more sustainable alternative. Amaranth grain contains 6 to 9% of oil which is higher than most cereal sources. Amaranth oil contains approximately 77% unsaturated fatty acids including linoleic acid. Additionally, the oil fraction is unique due to its high squalene content (2.4–8.0%). Since the clinical and nutraceutical applications of squalene are continuously growing, the demand for this substance will likely increase and amaranth oil could be a good alternative source [14,15]. One of the disadvantages of amaranth oil is its high susceptibility to oxidation, so the preparation of Pickering emulsions would be a good method to protect the oil and its valuable squalene content.

Previous studies have shown that whey protein isolate nanoparticles [5] and nanofibrils [7] can be used as stabilizers in Pickering emulsion systems to improve the stability of lipophilic bioactive compounds. In this study, the use of nanoparticles of one of the proteins (α-lactalbumin) from bovine whey is proposed as a stabilizer for Pickering emulsions containing squalene-rich amaranth oil. Bovine α-lactalbumin (α-LA) constitutes around 20 to 25% of the whey proteins, being the second most abundant protein in whey. It has a molecular weight of 14,178 Da, its isoelectric point is 4.2 to 4.6, it is very soluble in water and NaCl solutions, and it is relatively thermostable when bound to calcium [16]. It is the regulatory element in the lactose synthase complex, and it is involved in the modulation of the specificity of the galactosyl transferase subunit [17]. α-LA has been used previously to encapsulate lipophilic bioactive compounds such as retinol, α-tocopherol, and vitamin D_3 [18,19]; however, the preparation of α-LA nanostructures as possible carriers for drug delivery and food applications has acquired great interest [20]. Protein nanostructures have been primarily used for the encapsulation and controlled release of bioactive compounds in food systems. They offer numerous advantages including a lack of toxicity, in vivo degradation, and aqueous solubility and stability as in the case of α-LA [20]. These nanostructures can be prepared by the self-assembly of α-LA hydrolysates or by different physical or chemical treatments on the native α-LA protein. The controlled hydrolysis of the α-LA can produce by spontaneous assembly nanotubes [21], nanodisks [22], and nanospheres [23], which can be utilized for the delivery of bioactive compounds [24]. Self-assembled α-LA-derived peptides have been used as Pickering stabilizers for oil-in-water emulsions of curcumin [23]. When the native α-LA is to be used as the raw material for nanoparticle formation, the methods of desolvation with ethanol or acetone followed by chemical cross-linking [25,26] or enzymatic crosslinking with microbial transglutaminase or horseradish peroxidase [27,28] have been used. However, the preparation of Pickering emulsions stabilized by native α-LA nanoparticles has not been explored to the better of our knowledge. The aim of this study was, then, to study the conditions to prepare squalene-rich amaranth oil Pickering emulsions stabilized by α-LA nanoparticles. To accomplish this, α-LA nanoparticles, prepared at pH 9 and 11 using a desolvation and cross-linking process, were added at different concentrations to stabilize amaranth oil Pickering emulsions developed by ultrasound and rotor-stator homogenization. The characteristics, stability, and microstructure of the oil-in-water Pickering emulsions were studied to define the preparation conditions for maximal stability.

2. Materials and Methods

2.1. Materials

α-LA (BiPRO® Alpha 9000) was generously provided by Agropur Ingredients (Eden Prairie, MN, USA), amaranth oil was purchased from Gastronomía Molecular (Mexico City, Mexico), and all the other chemicals were bought from Sigma-Aldrich (St. Louis, MO, USA) and J.T. Baker (Mexico City, Mexico). They were of analytical grade and used as obtained.

2.2. Preparation of Native α-LA Nanoparticles

The α-LA nanoparticles were prepared according to the desolvation and cross-linking method described by Arroyo–Maya et al. [25]. Briefly, the α-LA (40 mg) was dissolved in 2 mL of a 10 mM NaCl solution (pH 9) or a 10 mM NaOH solution (pH 11) followed by the desolvation of the protein in solution by the controlled (1 mL/min) dropwise addition of 8 mL of acetone with constant mixing (500 rpm). After the desolvation step, 40 μL of 8% aqueous glutaraldehyde solution was added to attain the cross-linking of the α-LA. After a 3 h stirring step, the resultant nanoparticles were purified by five cycles of centrifugation (11500 rpm, 30 min, 4 °C). The final pellet was redispersed in the original volume of a 10 mM NaCl solution (pH 9.0) or a 10 mM NaOH solution (pH 11) using a Cole-Parmer 130-Watt Ultrasonic Processor. All preparations were performed at room temperature (25 °C).

2.3. Measurements of ζ Potential, Particle Size, and Size Distribution of the α-LA Nanoparticles

The ζ potential, particle size, and polydispersity index of the α-LA nanoparticles were measured immediately after preparation utilizing a Zetasizer Nano ZS90 (Malvern Instruments Inc., Malvern, U.K.) dynamic light scattering (DLS) instrument at 25 °C [26].

2.4. Microscopy Analysis of the α-LA Nanoparticles

The morphology of the α-LA nanoparticles was examined with a JEM1010 transmission electron microscope (JEOL, Tokyo, Japan) at 60 kV. The water suspension of the nanoparticles was diluted 10 times, and an aliquot of 5 μL was drop-casted onto a formvar carbon-coated copper grid (200 mesh); the grid was dried with air at room temperature before TEM imaging [25].

The morphometric characteristics of the nanoparticles were analyzed with a Quanta 200 FEG (FEI Company, Hillsboro, OR, USA) Dual Beam environmental scanning electron microscope using acceleration voltages of 10 and 20 kV for magnifications of 25,000 and 50,000 x respectively. The water suspension of the nanoparticles was diluted at 1:100 at 25 °C before observation [29].

The size and morphometric characteristics of the nanoparticles were also evaluated with a Nanosurf Naio Atomic Force Microscope (Liestal, Switzerland). The samples (10 μL) were allowed to dry at room temperature. The images were taken in dynamic force (tapping) mode with a Tap190Al-G probe [30]. All the observations were performed with freshly prepared nanoparticles.

2.5. Preparation of the Pickering Emulsions

Pickering o/w emulsions stabilized by the native α-LA nanoparticles obtained at pH 9 (NP1) and pH 11 (NP2) were prepared as described by Leal-Castañeda et al. [6] by mixing 17.5 mL of water with 7.5 mL of squalene oil and nanoparticles (NP1 and NP2) to reach concentrations of 0, 3, 5, 10, 15, and 20% wt. The mixtures were blended at 11,000 rpm for 3 min at 25 °C with an IKA T18 Basic Ultra-Turrax rotor-stator homogenizer followed by a 3 min ultrasound treatment in a Cole-Parmer 130-Watt Ultrasonic Processor at 25 °C.

2.6. ζ Potential, Particle Size, and Size Distribution of the Pickering Emulsions Droplets

The droplet size distribution, polydispersity index, and ζ potential of the emulsions (stabilized with NP1 and NP2 at concentrations of 0 to 20%) were measured using a Malvern Nano-S90 laser particle size analyzer 8 h after preparation at 25 °C [7].

2.7. Emulsion Stability

The stability of the emulsions with time was measured using a Turbiscan Lab Expert (Formulaction, Toulouse, France). The measurement principle of this instrument has been explained by Zequan et al. [31]. By scanning the sample at specific intervals, a pattern of the light flux (transmitted or backscattered) as a function of the sample height can be obtained, which is characteristic for each sample, so a parameter called Turbiscan Stability Index (TSI) can be calculated with the aid of the following equation:

$$TSI = \sqrt{\frac{\sum_{i=1}^{n}(x_i - x_{bs})^2}{n-1}} \tag{1}$$

where x_i is the average backscatter per minute of measurement, x_{bs} is the average value of x_i, and n is the number of scans [6]. The TSI includes all the variations occurring in the system including the most common instability mechanisms that occur in food emulsions (creaming, flocculation, sedimentation, coalescence, Ostwald ripening, and phase inversion) [32]. The higher the TSI value, the lower the stability of the emulsion. A sample (25 mL) of the Pickering emulsion was added to the equipment glass cell and allowed to remain for 8 h at 25 °C and the Global TSI was then calculated as the arithmetic mean of the individual TSI values.

2.8. Emulsification Efficiency

The emulsification efficiency (EE%) is defined by the concentration of the incorporated material (squalene oil) in the Pickering emulsion over the initial concentration used to make the formulation. The EE % was calculated using Equation (2):

$$EE\% = \frac{ml\ of\ encapsulated\ oil}{ml\ of\ total\ oil} \times 100 \tag{2}$$

The amount of encapsulated oil was calculated as the amount of total oil minus the amount of free oil which was released from the system by centrifugation (3000 rpm for 10 min) immediately after the Pickering emulsion was formed [33].

2.9. Confocal Laser Scanning Microscopy (CLSM)

A Confocal Laser Scanning Microscope (CLSM) (LSM 710 NL0 Carl Zeiss, Oberkochen, Germany) was employed to visualize the structure of the Pickering emulsion droplets 8 h after their preparation. Fluorescein isothiocyanate (FITC) was used to stain the α-LA nanoparticles [34], and Nile Red dye was used to stain the oil phase [6]. The freshly prepared Pickering emulsions (1 μL) were placed along with the dyes between two cover glasses in the CLSM and operated in fluorescence mode with an Apochromat-Plan objective (63×/1.4 Oil DIC M27) at 60×/1.4.

2.10. Statistical Analysis

The results were reported as the arithmetic mean ± standard deviation of three replicates. The Student t-test and one-way ANOVA were used to detect statistical differences among treatments (α = 0.05).

3. Results and Discussion

3.1. Preparation and Characterization of α-LA Nanoparticles

Dynamic light scattering is a powerful technique used to determine the size, size distribution, and surface charge (ζ potential) of nanoparticles, obtaining information about their stability and surface interaction with other molecules [35]. Table 1 shows the experimental values of these parameters for the α-LA nanoparticles prepared by the desolvation and cross-linking method from protein solutions with initial pH values of 9 (NP1) and 11 (NP2).

Table 1. Effect of pH of the initial protein solution on the α-LA nanoparticles size, size distribution, and ζ potential.

Nanoparticle	Size (nm)	Polydispersity Index (PDI)	ζ Potential (mV)
NP1 (pH 9)	143 ± 8 [a]	0.09 ± 0.01 [a]	−29.7 ± 1.42 [a]
NP2 (pH 11)	152 ± 1 [b]	0.06 ± 0.04 [b]	−25.4 ± 2.35 [b]

Values in a column followed by different lowercase superscript letters were significantly different (α = 0.05) from each other according to the Student *t*-test for paired measurements. Data are expressed as the mean of 36 replicates ± standard deviation.

It can be noted that the nanoparticles prepared at pH 11 (NP2) were significantly larger ($p \leq 0.05$) than the nanoparticles prepared at pH 9 (NP1). The PDI value of the NP1 was significantly larger than the PDI of the NP2. PDI is a representation of the size distribution within certain particle populations. Values of PDI ≤ 0.02 are considered adequate since the nanoparticle distributions can be considered monodisperse [36]. The size and size distribution of the α-LA nanoparticles fall in the range of the results obtained by other authors using the desolvation method for the same protein [20,25,26,37]. The ζ potential for both α-LA nanoparticles was near −30 mV (−29.7 and −25.4 mV), which indicated that the two α-LA nanoparticle systems had good stability [26]. In the case of α-LA nanoparticles prepared by desolvation and cross-linking, a negative surface electric charge at neutral and basic values of pH has been reported [38]. A minimal ζ potential of −30 mV is required for electrostatically stable nanoparticles and a minimal of −20 mV for steric stabilization [39]. Our data are consistent with the ζ potential values observed for other albumin nanoparticles [26,37]. These results indicate that, at the pH values used in the preparation (9 and 11) of the α-LA nanoparticles, electrostatic repulsion may play an important role in the Pickering stabilization of emulsions.

It can be seen (Figure 1) that the morphology of the α-LA nanoparticles, as observed by SEM, is basically spherical with particle sizes of approximately 150 nm. These sizes are consistent with the sizes determined by DLS in this study and by other authors [25,37,38] by TEM.

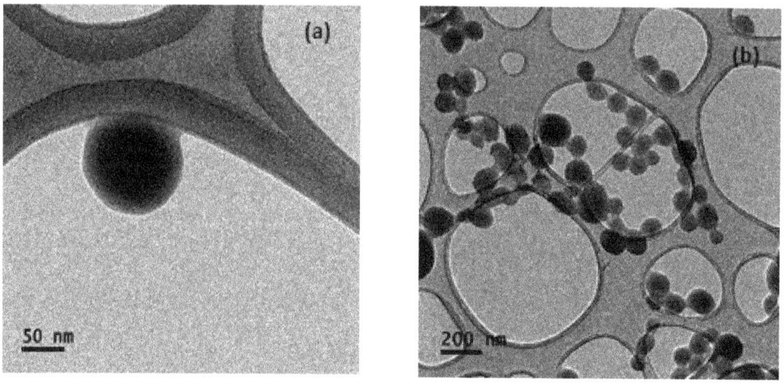

Figure 1. TEM images of α-LA nanoparticles formed from protein solutions with different pH values: (**a**) pH 7 and (**b**) pH 9.

SEM images were taken to evaluate the general and surface morphology of the nanoparticles. The nanoparticle size on the images matched, again, well the DLS data. Most of the α-LA nanoparticles have regular spherical shapes with average diameters of 150 nm (Figure 2) and are similar to the images of other SEM studies [26].

Atomic force microscopy (AFM) has been effectively applied in the field of food proteins. There are many applications (nanoimaging, force spectroscopy, manipulation, etc.) of AFM; however, nanoimaging is the most popular method to evaluate the structures, functions, sizes, and distributions in proteins [30,40,41]. In the case of both kinds of α-LA nanoparticles, semispherical surfaces can be observed (Figure 3) with particle sizes in the range of 150 to 165 nm. Nanoparticle agglomerates can be distinguished in all cases. The nanoparticle size and size distribution were, again, very similar to the data obtained by DLS. This similarity was previously observed by Hoo et al. [41], who also indicated that AFM has advantages over DLS just in the case of non-monodisperse solutions. This is the first time that native α-LA nanoparticles have been visualized by AFM. The only antecedents were the observations of disk-shaped nanoparticles and nanotubes obtained upon partial enzymatic hydrolysis of α-LA [22,42].

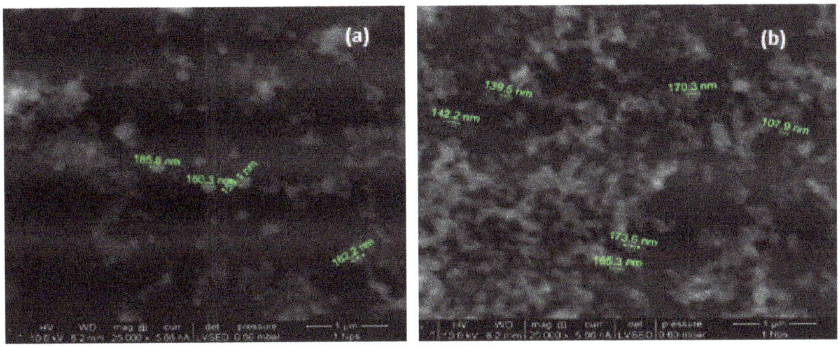

Figure 2. SEM images of α-LA nanoparticles formed from protein solutions with different pH values: (**a**) pH 7 and (**b**) pH 9.

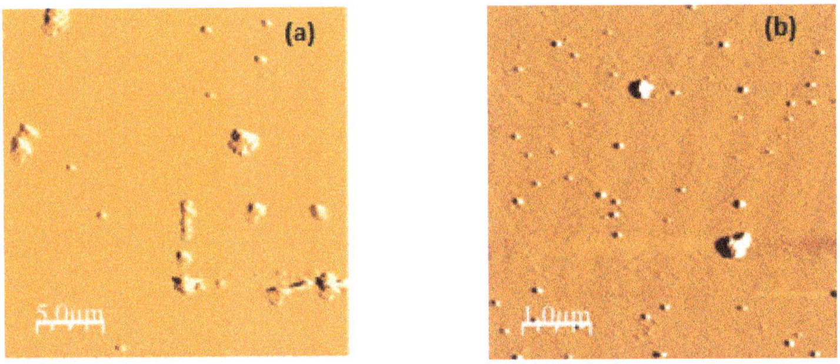

Figure 3. AFM images of α-LA nanoparticles formed from protein solutions with different pH values: (**a**) pH 7 and (**b**) pH 9.

3.2. Droplet Size, Polydispersity Index, and ζ Potential of the Pickering Emulsions

There are two main factors affecting the emulsion droplet size and size distribution during the preparation of Pickering emulsions. One of them is the concentration of α-LA nanoparticles initially dispersed in the aqueous phase and the other is the amount of amaranth oil phase. In this case, the second factor was maintained constant at 30% following the recommendation of Castañeda-Leal et al. [6]. The effect of the concentration of α-LA nanoparticles in the aqueous phase was observed by measuring the droplet size, polydispersity index, and ζ potential over a wide range of nanoparticle concentrations. The results can be observed in Table 2.

The Table shows that the average emulsion droplet size decreased from 5.8 ± 0.4 to 1.1 ± 0.5 μm and from 5.9 ± 0.2 to 2.2 ± 1.4 μm as NP1 and NP2 concentrations increased from 3 to 20% at a constant oil fraction (0.3). No significant difference between the two sets of droplet sizes (NP1 and NP2) could be detected according to the Student t-test for paired measurements ($p > 0.05$). This behavior has been previously described in the case of Pickering emulsions stabilized with soy protein isolate-chitosan [4] and starch [6] nanoparticles. This dependence of the droplet size on the concentration of stabilizing particles has been previously described and explained as a phenomenon in which larger amounts of solid particles are able to generate smaller droplets with limited coalescence due to the formation of an adsorbed layer of solid particles forming a stiff coating around the liquid droplets [43]. In the case of this study, the variation of the droplet size with the concentration of α-LA nanoparticles could be modeled with polynomial equations as follows:

$$D_{NP1} = -0.0028C^3 + 0.088C^2 - 0.9893C + 8.0672 \quad (3)$$

$$D_{NP2} = 0.0122C^2 - 0.4982C + 7.2848 \quad (4)$$

where D_{NP1} and D_{NP2} are the diameters of the droplets of the emulsions stabilized with NP1 and NP2 and C is the concentration (%) of NP1 and NP2 in equations 3 and 4, respectively. The determination coefficients (R^2) were 0.9998 and 1.0000, respectively. The average droplet size in this study was smaller than the sizes reported in other studies where the authors used other kinds of nanoparticles [4,6,8]. No significant difference in the droplet size was detected ($p \leq 0.05$) as a result of the pH values used in the preparation (9 and 11) of the α-LA nanoparticles (NP1 and NP2).

Table 2. Droplet size, polydispersity index, ζ potential, global turbiscan stability index, and encapsulation efficiency of emulsions at different concentrations of α-LA nanoparticles.

Concentration of α-LA Nanoparticles (%)	Droplet Size (μm)	PDI	ζ Potential	Global TSI	Emulsification Efficiency (%)
NP1-3	5.8 ± 0.4	0.73 ± 0.16	−59.8 ± 8.3	8.63 ± 0.7	40
NP1-5	5.0 ± 0.4	0.65 ± 0.17	−39.7 ± 1.95 ^{a}	6.46 ± 1.3	48
NP1-10	4.15 ± 0.75	0.42 ± 0.03 ^{a}	−44.7 ± 3.2 ^{a,b}	1.86 ± 0.07 ^{b}	92
NP1-15	3.6 ± 0.66	0.49 ± 0.10 ^{a}	−39.7 ± 15.8 ^{a,b}	1.98 ± 0.2 ^{b}	84
NP1-20	1.1 ± 0.5	0.74 ± 0.04 ^{b}	−40.6 ± 5.4	2.06 ± 0.7	87
NP2-3	5.9 ± 0.2	0.88 ± 0.12	−42.5 ± 3.4 ^{a,b}	4.7 ± 2.7	46
NP2-5	5.1 ± 0.37	0.61 ± 0.11	−41.7 ± 5.4 ^{a,b}	2.0 ± 0.7	51
NP2-10	3.52 ± 0.34	0.60 ± 0.16	−41.9 ± 20	1.6 ± 0.04 ^{a}	94
NP2-15	2.56 ± 0.36	0.66 ± 0.16	−33.4 ± 9.9	1.8 ± 0.8	86
NP2-20	2.2 ± 1.4	0.88 ± 0.15	−48.7 ± 3.7 ^{b}	1.7 ± 0.1 ^{a,b}	89

Different superscript lowercase letters within the same column indicate significant difference (α = 0.05). Only the samples with the smaller standard deviations were tested. Data are expressed as the mean of 3 replicates ± standard deviation. NP1 and NP2 are nanoparticles prepared at pH 9 and 11.

Statistically, it is difficult to work with small samples which come from populations with large standard deviations since, in these cases, the false positive rate is considerable when evaluating significant differences using One-way ANOVA [44]. However, adequate samples for the statistical evaluation of significant differences in PDI could be obtained in the cases of NP1-10, NP1-15, and NP1-20. In this case, the smallest PDI values (0.42 and 0.49) were found when 10 and 15% of NP1 initially dispersed in the aqueous phase were used (see Table 2). PDI is one of the important parameters to study the stability of emulsions. It is known that PDI values below 0.5 are indicative of the reliability of the measurements and the uniform distribution of the droplets [45]. In the case of this study, the smallest PDI values (0.42 and 0.60) were obtained for the Pickering emulsions prepared with 10% NP1 and NP2, respectively. These values are similar to those obtained in other works [46].

ζ potential is a critical parameter for assessing the stability of Pickering emulsions and for choosing adequate wall materials to be used as stabilizers. Generally, emulsion stability has been associated with ζ potential values of ± 30 mV [7]. In the case of this study, values of ζ potential between −39.7 and −59.8 mV and between −33.4 and −48.7 mV were obtained for the emulsions prepared with NP1 and NP2, respectively (see Table 2). All the values fall in the range of values reported by other authors [7,47] and in the interval of stability for emulsions (± 30 mV).

3.3. Confocal Laser Scanning Microscopy (CLSM)

In Figure 4a,b, images of the Pickering emulsions prepared with a 15% concentration of NP1 and NP2 are shown: red regions correspond to the oil phase and are surrounded by a white to green membrane (α-LA nanoparticles) on a light green background. Average droplet sizes between 3 and 4 μm can be observed along with some smaller droplets and individual agglomerates of nanoparticles stained in dark green. The thickness of the adsorbed layer of solid particles is between 700 and 1000 nm and depends on the size of the droplets to which they are adsorbed.

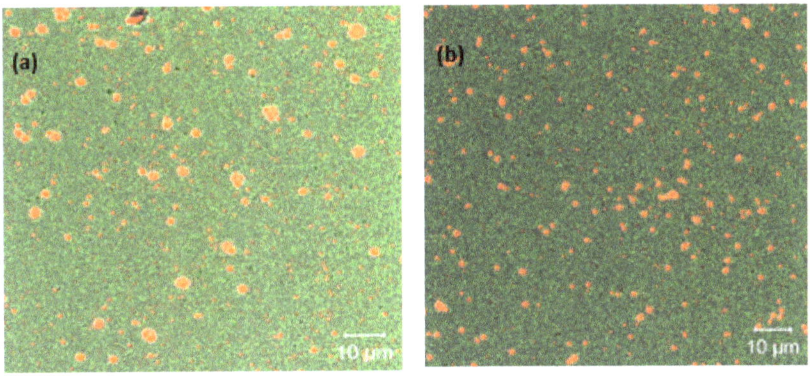

Figure 4. CLSM micrographs of amaranth oil Pickering emulsions stabilized with 15% NP1 (**a**) and NP2 (**b**).

3.4. Stability of Pickering Emulsions Determined by Light Backscattering

Emulsion stability can be defined as "the ability of an emulsion to resist changes in its physicochemical properties over time" [32]. The stability of the emulsions with time was measured as the Global Turbiscan Stability Index (TSI). The Global TSI after 8 h of storage at different concentrations of NP1 and NP2 is shown in Table 2. The photographs of the Pickering emulsions after 8 h of storage are shown in Figure 5.

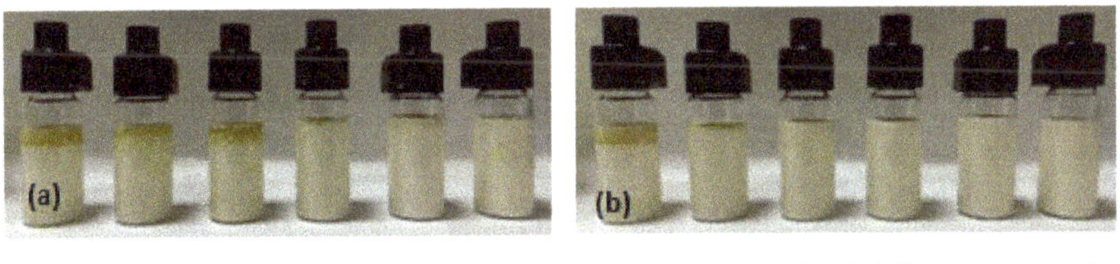

Figure 5. Photographs of the amaranth oil Pickering emulsions prepared with different concentrations (0, 3, 5, 10, 15, and 20%) of (**a**) NP1 and (**b**) NP2 after 8 h of storage at 25 °C.

The Global TSI decreased from 8.63 to 1.98 as NP1 concentration increased from 3 to 15%, indicating an increase in the stability of the Pickering emulsions. No oil separation can be observed for the emulsions with 10, 15, and 20% of NP1 (Figure 5a). This was to be expected since they have the lowest values of Global TSI (Table 2). In the case of the emulsions stabilized with NP2, the Global TSI decreased from 4.7 to 1.7 as the nanoparticle concentrations increased from 3 to 20%. The oil separation could not be observed in the samples with 5, 10, 15, and 20% of nanoparticles (Figure 5b). Again, these concentrations formed the emulsions with the lowest Global TSI values (Table 2). This indicates that 10 and 5% are the concentrations necessary for NP1 and NP2 to reach the critical coverage ratio. If this ratio is not achieved, coalescence is likely to occur. There are many instability mechanisms occurring in food emulsions, however, under certain conditions, Ostwald ripening is considered the main mechanism in the case of Pickering emulsions [48]. However, the mechanism is inversely proportional to the droplet size. For droplets larger than 1 μm, the process occurs only at a very slow rate, which is probably the case of the system in this study. Global TSI values of 1.86 for NP1-10% and 1.6 for NP2-10% are the amaranth oil Pickering emulsions with the highest stabilities with time. These values are comparable to those obtained by Leal–Castañeda et al. [6] for Pickering emulsions stabilized with amaranth grain starch after 6 h of storage. The emulsified region in all the studied systems was located on the right side of the Turbiscan graphs, indicating that the instability region is located at the top of the emulsion and confirming that the main instability mechanism is creaming. Previous reports indicate that in the case of Pickering emulsions, the storage stability can be measured by creaming since these systems show high stability against coalescence [49].

3.5. Emulsification Efficiency

The higher emulsification efficiencies were obtained in the cases of the NP1-10% and NP2-10% emulsions with 92 and 94%, respectively (see Table 2). There are reports that in the case of polysaccharides, such as chitosan, the ζ potential is a good indicator of the emulsification efficiency of colloidal particles [49]. In the case of this study, the α-LA nanoparticles have ζ potential values between −29.7 and −25.4 mV, which apparently could indicate that a good emulsification efficiency could be achieved once the critical coverage ratio is reached.

Considering the characteristics of the Pickering emulsions prepared with different concentrations of NP1 and NP2 as stabilizers, the formulations with the best functionality were those containing NP1-10% and NP2-10% (see Table 2). These formulations have a good droplet size (4.15 and 3.52 μm), they are relatively homogeneous (PDI values of 0.42 and 0.60), they have a ζ potential indicative of high stability (−44.7 and −41.9 mV) which was corroborated by the lowest values of Global TSI (1.86 and 1.60), and they have the highest values of emulsification efficiency (92 and 98%). These formulations are, therefore, the best choices for the preparation of amaranth oil Pickering emulsions for the adequate delivery of squalene. Actually, they could have wide application possibilities as food delivery systems for lipophilic bioactive compounds. A previous study considered the possibility of preparing nanoparticle-stabilized Pickering emulsions to transform liquid oils into soft solids that could be used as a trans-fat-free replacement for partially hydrogenated oils utilized in margarines, shortenings, and bakery products [50].

4. Conclusions

Food grade squalene-rich amaranth oil Pickering emulsions stabilized by α-LA nanoparticles were successfully prepared. It was found that α-LA nanoparticles prepared by the desolvation and cross-linking method from protein solutions with pH values of 9 (NP1) and 11 (NP2) acted as good stabilizers for Pickering emulsions. Dependence of the droplet size of the emulsions on the nanoparticle concentration was observed. The critical coverage ratio was reached when 10 and 5% nanoparticle concentrations were used for NP1 and NP2, respectively. Our findings suggest that the NP1-10% and NP2-10% can be used as novel stabilizers for Pickering emulsions to provide protection for valuable lipophilic bioactive compounds such as squalene. This is the first time that native α-LA nanoparticles have been used as stabilizers of Pickering emulsions.

Author Contributions: Conceptualization, H.H.-S. and I.J.A.-M.; methodology, A.P.C.-G., I.J.A.-M., G.F.G.-L. and H.H.-S.; software, H.H.-S.; validation, A.P.C.-G., I.J.A.-M. and H.H.-S.; formal analysis, A.P.C.-G., B.G.-M., I.J.A.-M. and M.C.-M.; investigation, A.P.C.-G., I.J.A.-M. and H.H.-S.; resources, I.J.A.-M., G.F.G.-L. and H.H.-S.; data curation, A.P.C.-G., I.J.A.-M. and H.H.-S.; writing—original draft preparation, A.P.C.-G., I.J.A.-M. and H.H.-S.; writing—review and editing, I.J.A.-M. and H.H.-S.;

visualization, H.H.-S.; supervision, H.H.-S.; project administration, H.H.-S.; funding acquisition, G.F.G.-L. and H.H.-S. All authors have read and agreed to the published version of the manuscript.

Funding: This research was funded by SIP-Instituto Politécnico Nacional (Mexico), grant numbers 20210460 and 20221192.

Data Availability Statement: The data presented in this study are available on request from the corresponding authors.

Acknowledgments: Authors A.P.C.G. and B.G.M. acknowledge CONACyT for their scholarships. The authors appreciate the help provided by Centro de Nanociencias y Micro y Nanotecnologías (IPN) with the electronic microscopy experiments.

Conflicts of Interest: The authors declare no conflict of interest. The funders had no role in the design of the study; in the collection, analyses, or interpretation of data; in the writing of the manuscript, or in the decision to publish the results.

References

1. Binks, B.P.; Dezhong, Y. Pickering emulsions stabilized by hydrophilic nanoparticles: In situ surface modification by oil. *Soft Matter* **2016**, *12*, 6858–6867. [CrossRef] [PubMed]
2. Ridel, L.; Bolzinger, M.A.; Gilon-Delepine, N.; Dugas, P.Y.; Chevalier, Y. Pickering emulsions stabilized by charged nanoparticles. *Soft Matter* **2016**, *12*, 7564–7576. [CrossRef] [PubMed]
3. De Folter, J.W.J.; van Ruijven, M.W.M.; Velikov, K.P. Oil-in-water Pickering emulsions stabilized by colloidal particles from the water-insoluble protein zein. *Soft Matter* **2012**, *8*, 6807–6815. [CrossRef]
4. Yang, H.; Su, Z.; Meng, X.; Zhang, X.; Kennedy, J.F.; Liu, B. Fabrication and characterization of Pickering emulsions stabilized by soy protein isolate-chitosan nanoparticles. *Carbohydr. Polym.* **2020**, *247*, 116712. [CrossRef] [PubMed]
5. Wu, J.; Shi, M.; Li, W.; Zhao, L.; Wang, Z.; Yan, X. Pickering emulsions stabilized by whey protein nanoparticles prepared by thermal cross-linking. *Colloids Surf. B* **2015**, *127*, 96–104. [CrossRef]
6. Leal-Castañeda, E.J.; Farcía-Tejeda, Y.; Hernández-Sánchez, H.; Alamilla-Beltrán, L.; Téllez-Medina, D.I.; Calderón-Domínguez, G.; García, H.S.; Gutiérrez-López, G.F. Pickering emulsions stabilized with native and lauroylated amaranth starch. *Food Hydrocoll.* **2018**, *80*, 177–185. [CrossRef]
7. Jiao, Q.; Liu, Z.; Li, B.; Tian, B.; Zhang, N.; Liu, C.; Feng, Z.; Jiang, B. Development of antioxidant and stable conjugated linoleic acid Pickering emulsion with protein nanofibers by microwave-assisted self-assembly. *Foods* **2021**, *10*, 1892. [CrossRef]
8. Wu, X.; Li, X.; Yang, L.; Yuan, L.; Xu, Z.; Ge, L.; Mu, C.; Li, D. Stability enhanced Pickering emulsions based on gelatin and dialdehyde starch nanoparticles as simple strategy for structuring liquid oils. *Food Bioprocess Technol.* **2021**, *14*, 1600–1610. [CrossRef]
9. Yan, X.; Ma, C.; Cui, F.; McClements, D.J.; Liu, X. Protein-stabilized Pickering emulsions: Formation, stability, properties, and applications in foods. *Trends Food Sci. Technol.* **2020**, *103*, 293–303. [CrossRef]
10. Huang, Z.R.; Lin, Y.K.; Fang, J.Y. Biological and pharmacological activities of squalene and related compounds: Potential uses in cosmetic dermatology. *Molecules* **2009**, *14*, 540–554. [CrossRef]
11. Mendes, A.; Azevedo-Silva, J.; Fernandes, J.C. From sharks to yeasts: Squalene in the development of vaccine adjuvants. *Pharmaceuticals* **2022**, *15*, 265. [CrossRef] [PubMed]
12. Kelly, G.S. Squalene and its potential clinical uses. *Altern. Med. Rev.* **1999**, *4*, 29–36. [PubMed]
13. Micera, M.; Botto, A.; Geddo, F.; Antoniotti, S.; Bertea, C.M.; Levi, R.; Gallo, M.P.; Querio, G. Squalene: More than a step toward sterols. *Antioxidants* **2020**, *9*, 688. [CrossRef] [PubMed]
14. He, H.P.; Corke, H. Oil and squalene in *Amaranthus* grain and leaf. *J. Agric. Food Chem.* **2003**, *51*, 7913–7920. [CrossRef]
15. Martirosyan, D.M.; Miroshnichenko, L.A.; Kulakova, S.N.; Pogojeva, A.V.; Zoloedov, V.I. Amaranth oil application for coronary heart disease and hypertension. *Lipids Health Dis.* **2007**, *6*, 1. [CrossRef]
16. Kamau, S.M.; Cheison, S.C.; Chen, W.; Liu, X.M.; Lu, R.R. Alpha-Lactalbumin: Its production technologies and bioactive peptides. *Compr. Rev. Food Sci. Food Saf.* **2010**, *9*, 197–211. [CrossRef]
17. Permyakov, E.A. α-Lactalbumin, amazing calcium-binding protein. *Biomolecules* **2020**, *10*, 1210. [CrossRef]
18. Nareswara, A.R.; Alamsyah, A.Z.; Afifah, D.N.; Panunggal, B.; Sulchan, M.; Khumaeni, A.; Anjani, G. Encapsulation efficiency of vitamin D_3 in α-lactalbumin during storage. *Food Res.* **2020**, *4* (Suppl. S3), 141–146. [CrossRef]
19. Chen, W.; Yu, H.; Shi, R.; Ma, C.; Gantumur, M.A.; Qayum, A.; Bilawal, A.; Liang, G.; Oh, K.C.; Jiang, Z. Comparison of carrying mechanisms between three fat-soluble vitamins and alpha-lactalbumin: Effects on structure and physicochemical properties of alpha-lactalbumin. *Food Hydrocoll.* **2021**, *116*, 106662. [CrossRef]
20. Mehravar, R.; Jahanshani, M.; Saghatoleslami, N. Production of biological nanoparticles from α-lactalbumin for drug delivery and food science application. *Afr. J. Biotechnol.* **2009**, *8*, 6822–6827.
21. Graveland-Bikker, J.F.; de Kruif, C.G. Unique milk protein based nanotubes: Food and nanotechnology meet. *Trends Food Sci. Technol.* **2006**, *17*, 196–203. [CrossRef]

22. Balandrán-Quintana, R.R.; Valdéz-Covarrubias, M.A.; Mendoza-Wilson, A.M.; Sotelo-Mundo, R.R. α-Lactalbumin hydrolysate spontaneously produces disk-shaped nanoparticles. *Int. Dairy J.* **2013**, *32*, 133–135. [CrossRef]
23. Liu, B.; Liu, B.; Wang, R.; Li, Y. α-Lactalbumin self-assembled nanoparticles with various morphologies, stiffnesses, and sizes as Pickering stabilizers for oil-in-water emulsions and delivery of curcumin. *J. Agric. Food Chem.* **2021**, *69*, 2485–2492. [CrossRef]
24. Fuciños, C.; Míguez, M.; Fuciños, P.; Pastrana, L.M.; Rúa, M.L.; Vicente, A.A. Creating functional nanostructures: Encapsulation of caffeine into α-lactalbumin nanotubes. *Innov. Food Sci. Emerg. Technol.* **2017**, *40*, 10–17. [CrossRef]
25. Arroyo-Maya, I.J.; Hernández-Sánchez, H.; Jiménez-Cruz, E.; Camarillo-Cadena, M.; Hernández-Arana, A. α-Lactalbumin nanoparticles prepared by desolvation and cross-linking: Structure and stability of the assembled protein. *Biophys. Chem.* **2014**, *193–194*, 27–34. [CrossRef] [PubMed]
26. Etorki, A.M.; Gao, M.; Sadeghi, R.; Maldonado-Mejía, L.F.; Kokini, J.L. Effects of desolvating agent types, ratios, and temperature on size and nanostructure of nanoparticles from α-lactalbumin and ovalbumin. *J. Food Sci.* **2016**, *81*, E2511–E2520. [CrossRef]
27. Dhayal, S.K.; Gruppen, H.; de Vries, R.; Wierenga, P.A. Controlled formation of protein nanoparticles by enzymatic cross-linking of α-lactalbumin with horseradish peroxidase. *Food Hydrocoll.* **2014**, *36*, 53–59. [CrossRef]
28. Dhayal, S.K.; Delahaije, R.J.; de Vries, R.; Gruppen, H.; Wierenga, P.A. Enzymatic cross-linking of α-lactalbumin to produce nanoparticles with increased foam stability. *Soft Matter* **2015**, *11*, 7888–7898. [CrossRef]
29. Nudelman, R.; Gavriely, S.; Bychenko, D.; Barzilay, M.; Gulakhmedova, T.; Gazit, E.; Richter, S. Bio-assisted synthesis of bimetallic nanoparticles featuring antibacterial and photothermal properties for the removal of biofilms. *J. Nanobiotechnol.* **2021**, *19*, 452. [CrossRef]
30. Grobelny, J.; Del Rio, F.W.; Pradeep, N.; Kim, D.I.; Hackley, V.A.; Cook, R.F. Size measurement of nanoparticles using Atomic Force Microscopy. In *Characterization of Nanoparticles Intended for Drug Delivery*; McNeil, S.A., Ed.; Springer Science+Business Media, LLC: New York, NY, USA, 2011; pp. 71–82.
31. Liu, Z.; Yang, X.; Zhang, Q. Turbiscan: History, development, applications to colloids and dispersions. *Adv. Mater. Res.* **2014**, *936*, 1592–1596. [CrossRef]
32. McClements, D.J. Critical review of techniques and methodologies for characterization of emulsion stability. *Crit. Rev. Food Sci. Nutr.* **2007**, *47*, 611–649. [CrossRef] [PubMed]
33. Lopera, S.M.; Guzmán, C.; Cataño, C.; Gallardo, C. Development and characterization of folic acid microparticles formed by spray drying with gum arabic and maltodextrin. *Vitae* **2009**, *16*, 55–65.
34. Chaganti, L.K.; Venkatakrishnan, N.; Bose, K. An efficient method for FITC labelling of proteins using tandem affinity purification. *Biosci. Rep.* **2018**, *38*, BSR20181764. [CrossRef] [PubMed]
35. Carvalho, P.M.; Felicio, M.R.; Santos, N.C.; Gonçalves, S.; Domingues. Application of Light Scattering Techniques to nanoparticle characterization and development. *Front. Chem.* **2018**, *6*, 237. [CrossRef]
36. Danaei, M.; Dehghankhold, M.; Ataei, S.; Hasanzadeh Davarani, F.; Javanmard, R.; Dokhani, A.; Khorasani, S.; Mozafari, R.F. Impact of particle size and Polydispersity Index on the clinical applications of lipidic nanocarrier systems. *Pharmaceutics* **2018**, *10*, 57. [CrossRef] [PubMed]
37. Jiménez-Cruz, E.; Cuevas-Gómez, A.P.; Unsworth, L.; Cornejo-Mazón, M.; Arroyo-Maya, I.J.; Hernández-Sánchez, H. Poly-L-lysine-coated α-lactalbumin nanoparticles: Preparation, effect of pH, and stability under in vitro simulated gastrointestinal conditions. *J. Chem. Technol. Biotechnol.* **2022**, *97*, 1597–1603. [CrossRef]
38. Arroyo-Maya, I.J.; Rodiles-López, J.O.; Cornejo-Mazón, M.; Gutiérrez-López, G.F.; Hernández-Arana, A.; Toledo-Núñez, C.; Barbosa-Cánovas, G.V.; Flores-Flores, J.O.; Hernández-Sánchez, H. Effect of different treatments on the ability of α-lactalbumin to form nanoparticles. *J. Dairy Sci.* **2012**, *95*, 6204–6214. [CrossRef]
39. Patel, V.R.; Agrawal, Y.K. Nanosuspension: An approach to enhance solubility of drugs. *J. Adv. Pharm. Technol. Res.* **2011**, *2*, 81–87.
40. Shi, C.; He, Y.; Ding, M.; Wang, Y.; Zhong, J. Nanoimaging of food proteins by atomic force microscopy. Part II: Application for food proteins from different sources. *Trends Food Sci. Technol.* **2019**, *87*, 14–25. [CrossRef]
41. Hoo, C.M.; Starostin, N.; West, P.; Mecartney, M.L. A comparison of atomic force microscopy (AFM) and dynamic light scattering (DLS) methods to characterize nanoparticle size distributions. *J. Nanopart. Res.* **2008**, *10*, 89–96. [CrossRef]
42. Graveland-Bikker, J.F.; Schaap, I.A.T.; Schimdt, C.F.; de Kruif, C.G. Structural and mechanical study of a self-assembling protein nanotube. *Nano Lett.* **2006**, *6*, 616–621. [CrossRef] [PubMed]
43. Chevalier, Y.; Bolzinger, M.A. Emulsions stabilized with solid nanoparticles: Pickering emulsions. *Colloids Surf. A* **2013**, *439*, 23–34. [CrossRef]
44. Glass, G.V.; Peckham, P.D.; Sanders, J.R. Consequences of failure to meet assumptions underlying fixed effects analyses of variance and covariance. *Rev. Educ. Res.* **1972**, *42*, 237–288. [CrossRef]
45. Belgheisi, S.; Motamedzadegan, A.; Milani, J.M.; Rashidi, L.; Rafe, A. Impact of ultrasound processing parameters on physical characteristics of lycopene emulsion. *J. Food Sci. Technol.* **2021**, *58*, 484–493. [CrossRef] [PubMed]
46. Ngwabebhoh, F.A.; Erdagi, S.I.; Yildiz, U. Pickering emulsions stabilized nanocellulosic-based nanoparticles for coumarin and curcumin nanoencapsulations: In vitro release, anticancer and antimicrobial activities. *Carbohydr. Polym.* **2018**, *201*, 317–328. [CrossRef]
47. Liu, C.; Xu, Y.; Xia, W.; Jiang, Q. Enhancement of storage stability of surimi particles stabilized novel Pickering emulsions: Effect of different sequential ultrasonic processes. *Ultrason. Sonochem.* **2021**, *79*, 105802. [CrossRef]

48. Gauthier, G.; Capron, I. Pickering nanoemulsions: An overview of manufacturing processes, formulations, and applications. *JCIS Open* **2021**, *4*, 100036. [CrossRef]
49. Rayner, M.; Marku, D.; Eriksson, M.; Dejmek, P.; Wahlgren, M. Biomass-based particles for the formulation of Pickering type emulsions in food and topical applications. *Colloids Surf. A* **2014**, *458*, 48–62. [CrossRef]
50. Zeng, T.; Wu, Z.L.; Zhu, J.Y.; Yin, S.W.; Tang, C.H.; Wu, L.Y.; Yang, X.Q. Development of antioxidant Pickering high internal phase emulsions (HIPEs) stabilized by protein/polysaccharide hybrid particles as potential alternative for PHOs. *Food Chem.* **2017**, *231*, 122–130. [CrossRef]

MDPI AG
Grosspeteranlage 5
4052 Basel
Switzerland
Tel.: +41 61 683 77 34

Foods Editorial Office
E-mail: foods@mdpi.com
www.mdpi.com/journal/foods

www.ingramcontent.com/pod-product-compliance
Lightning Source LLC
LaVergne TN
LVHW070445061126
851858LV00011B/162